ANSWERS

ZOOMING IN

Precalculus Explorations With Technology

By

Stephen Armstrong, Ed.D.

LeTourneau University
Longview, Texas

Edited by

Lin McMullin

D&S Marketing Systems, Inc. Brooklyn, New York

ISBN 1-878621-58-0

Printed in the U.S.A.

Chapter 1

Exploration 1: How Much Do You Really Pay?

1. Monthly payment = 439.58
2. 6. and 7.

Month	Unpaid Balance	Payment	Amt Interest	Amount on Principal	Ratio for Interest	Ratio for Principal
1	$5,000.00	$ 489.58	$ 41.67	$ 447.91	0.91042	1.008333
2	$4,552.09	$ 489.58	$ 37.93	$ 451.65	0.90078	1.008333
3	$4,100.44	$ 489.58	$ 34.17	$ 455.41	0.88894	1.008333
4	$3,645.03	$ 489.58	$ 30.38	$ 459.20	0.87402	1.008333
5	$3,185.83	$ 489.58	$ 26.55	$ 463.03	0.85466	1.008333
6	$2,722.79	$ 489.58	$ 22.69	$ 466.89	0.82853	1.008333
7	$2,255.90	$ 489.58	$ 18.80	$ 470.78	0.79131	1.008333
8	$1,785.12	$ 489.58	$ 14.88	$ 474.70	0.73408	1.008333
9	$1,310.42	$ 489.58	$ 10.92	$ 478.66	0.63473	1.008333
10	$ 831.76	$ 489.58	$ 6.93	$ 482.65	0.41973	0.723326
11	$ 349.11	$ 352.02	$ 2.91	$ 349.11		
Totals		$ 5,247.82	$ 247.82	$ 5,000.00		

4.

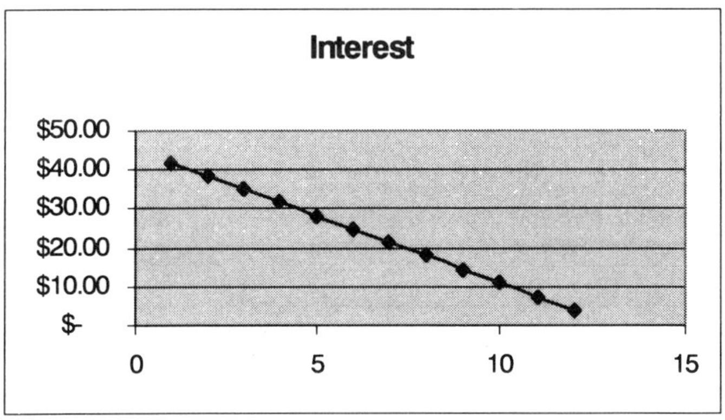

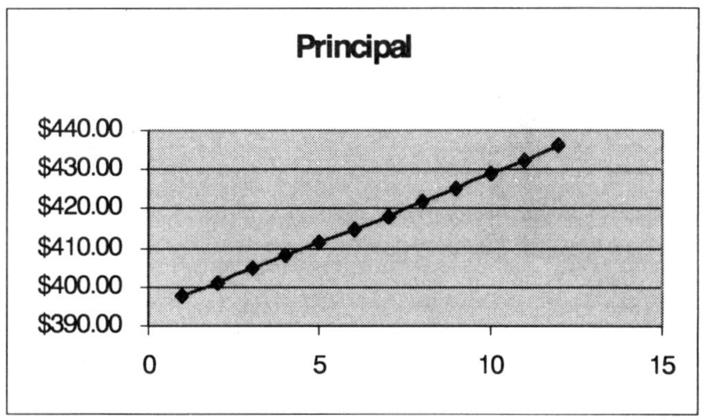

Graphs possibly appear linear but are not.

Exploration 2: Paying Off the Mortgage

1. Monthly payment $917.21
2.

Month	Remaining Principle	Amount of Interest	Amount Paid on Principal
1	$ 125,000.00	$ 833.33	$ 83.88
2	$ 124,916.12	$ 832.77	$ 84.44
3	$ 124,831.69	$ 832.21	$ 85.00
4	$ 124,746.69	$ 831.64	$ 85.57
5	$ 124,661.12	$ 831.07	$ 86.14

6	$ 124,574.99	$ 830.50	$ 86.71
7	$ 124,488.28	$ 829.92	$ 87.29
8	$ 124,400.99	$ 829.34	$ 87.87
9	$ 124,313.12	$ 828.75	$ 88.46
10	$ 124,224.66	$ 828.16	$ 89.05
	...	...	...
355	$ 5,370.97	$ 35.81	$ 881.40
356	$ 4,489.56	$ 29.93	$ 887.28
357	$ 3,602.28	$ 24.02	$ 893.19
358	$ 2,709.09	$ 18.06	$ 899.15
359	$ 1,809.94	$ 12.07	$ 905.14
360	$ 904.80	$ 6.03	$ 911.18
Totals		$205,189.22	$125,006.38

3. Total pay off = $ 330,195.60 Interest = $205,189.22

4. Extra $200 first 20 months saves 39 months, total interest = $173,201.02
Interest saved = $31,988.20

321	$ 808.99	$ 5.39	$ 911.82
322	$ (102.82)	$ (0.69)	$ 917.90

5. Extra $200 starting on 240th month saves 30 months, total interest = $195,820
Interest saved = $9,369.17

334	$ 167.11	$ 1.11	$ 1,116.10
335	$ (948.99)	$ (6.33)	$ 1,123.54

6. 298 months, Total interest = $147,313.19, a savings of $57,876.03

Exploration 3: Plastic Money

1.

Month	Current balance	Interest	Amt on Principal
1	$15,750.00	$ 137.81	$ (2.81)
2	$15,752.81	$ 137.84	$ (2.84)
3	$15,755.65	$ 137.86	$ (2.86)
4	$15,758.51	$ 137.89	$ (2.89)

2. No, each month the interest is more than the minimum payment. The principal keeps increasing.

3. When Payment is $150

286	$ 462.03	$ 4.04	$ 145.96
287	$ 316.07	$ 2.77	$ 147.23
288	$ 168.83	$ 1.48	$ 148.52
289	$ 20.31	$ 0.18	$ 149.82
290	$ (129.51)	$ (1.13)	$ 151.13

Total interest paid = $ 27,470.49

3a. Yes, b. 290, months, more than 24 years c. $ 27,470.49

4. When payment is $350

56	$ 843.23	$ 7.38	$ 342.62
57	$ 500.61	$ 4.38	$ 345.62
58	$ 154.99	$ 1.36	$ 348.64
59	$ (193.66)	$ (1.69)	$ 351.69

Total interest paid is $ 4,356.34

5. Payment is $511.91; Interest is $2678.91

6. To pay off in two years, payment is approximately $732

23	$ 1,403.81	$ 12.28	$ 719.72
24	$ 684.09	$ 5.99	$ 726.01
25	$ (41.92)	$ (0.37)	$ 732.37

Interest is approximately $1776

Exploration 4: Upside Down

1. First 24 Months of Payments:
2. & 3.

Payment Month	Payment	Interest	Amount on Principal	Remaining Balance	Car Value
1	280.53	$ 106.67	$ 173.86	$ 15,826.14	$15,733.34
2	280.53	$ 105.51	$ 175.02	$ 15,651.12	$15,471.12
3	280.53	$ 104.34	$ 176.19	$ 15,474.93	$15,213.27
4	280.53	$ 103.17	$ 177.36	$ 15,297.57	$14,959.71
5	280.53	$ 101.98	$ 178.55	$ 15,119.02	$14,710.38
6	280.53	$ 100.79	$ 179.74	$ 14,939.28	$14,465.21
7	280.53	$ 99.60	$ 180.93	$ 14,758.35	$14,224.12
8	280.53	$ 98.39	$ 182.14	$ 14,576.21	$13,987.05
9	280.53	$ 97.17	$ 183.36	$ 14,392.85	$13,753.94
10	280.53	$ 95.95	$ 184.58	$ 14,208.27	$13,524.70
11	280.53	$ 94.72	$ 185.81	$ 14,022.46	$13,299.29
12	280.53	$ 93.48	$ 187.05	$ 13,835.41	$13,077.64
13	280.53	$ 92.24	$ 188.29	$ 13,647.12	$12,859.68
14	280.53	$ 90.98	$ 189.55	$ 13,457.57	$12,645.35
15	280.53	$ 89.72	$ 190.81	$ 13,266.76	$12,434.59
16	280.53	$ 88.45	$ 192.08	$ 13,074.68	$12,227.35
17	280.53	$ 87.16	$ 193.37	$ 12,881.31	$12,023.56
18	280.53	$ 85.88	$ 194.65	$ 12,686.66	$11,823.17
19	280.53	$ 84.58	$ 195.95	$ 12,490.71	$11,626.12
20	280.53	$ 83.27	$ 197.26	$ 12,293.45	$11,432.35
21	280.53	$ 81.96	$ 198.57	$ 12,094.88	$11,241.81
22	280.53	$ 80.63	$ 199.90	$ 11,894.98	$11,054.44
23	280.53	$ 79.30	$ 201.23	$ 11,693.75	$10,870.20

24	280.53	$ 77.96	$ 202.57	$ 11,491.18	$10,689.03
25	280.53	$ 76.61	$ 203.92	$ 11,287.26	$10,510.88
26	280.53	$ 75.25	$ 205.28	$ 11,081.98	$10,335.70
27	280.53	$ 73.88	$ 206.65	$ 10,875.33	$10,163.44
28	280.53	$ 72.50	$ 208.03	$ 10,667.30	$ 9,994.05
29	280.53	$ 71.12	$ 209.41	$ 10,457.89	$ 9,827.48
30	280.53	$ 69.72	$ 210.81	$ 10,247.08	$ 9,663.69
31	280.53	$ 68.31	$ 212.22	$ 10,034.86	$ 9,502.63
32	280.53	$ 66.90	$ 213.63	$ 9,821.23	$ 9,344.25
33	280.53	$ 65.47	$ 215.06	$ 9,606.17	$ 9,188.51
34	280.53	$ 64.04	$ 216.49	$ 9,389.68	$ 9,035.37
35	280.53	$ 62.60	$ 217.93	$ 9,171.75	$ 8,884.78
36	280.53	$ 61.15	$ 219.39	$ 8,952.37	$ 8,736.70
37	280.53	$ 59.68	$ 220.85	$ 8,731.52	$ 8,591.09
38	280.53	$ 58.21	$ 222.32	$ 8,509.20	$ 8,447.91
39	280.53	$ 56.73	$ 223.80	$ 8,285.40	$ 8,307.11
40	280.53	$ 55.24	$ 225.29	$ 8,060.11	$ 8,168.66
41	280.53	$ 53.73	$ 226.80	$ 7,833.31	$ 8,032.51
42	280.53	$ 52.22	$ 228.31	$ 7,605.00	$ 7,898.64
43	280.53	$ 50.70	$ 229.83	$ 7,375.17	$ 7,766.99
44	280.53	$ 49.17	$ 231.36	$ 7,143.81	$ 7,637.54
45	280.53	$ 47.63	$ 232.90	$ 6,910.91	$ 7,510.25
46	280.53	$ 46.07	$ 234.46	$ 6,676.45	$ 7,385.08
47	280.53	$ 44.51	$ 236.02	$ 6,440.43	$ 7,262.00
48	280.53	$ 42.94	$ 237.59	$ 6,202.84	$ 7,140.96
49	280.53	$ 41.35	$ 239.18	$ 5,963.66	$ 7,021.95
50	280.53	$ 39.76	$ 240.77	$ 5,722.89	$ 6,904.91

5.

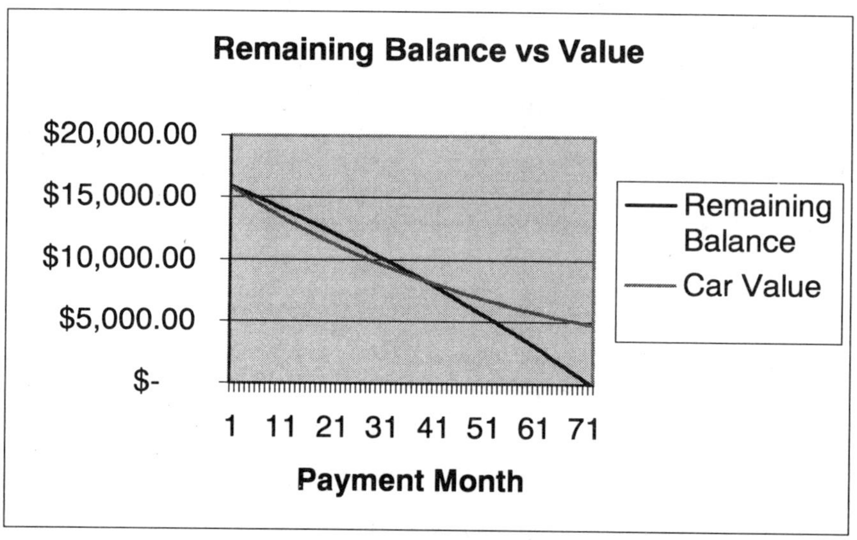

6. Upside down for 39 Months. Greatest difference is for the 20th month at $864.59

7. For the first 45 months, you would have to pay extra to settle the loan, even with $500 deductible.

8.

Payment Month	36 Month Loan Payment	Interest	Amount on Principal	Remaining Balance	Car Value
1	501.38	$ 106.67	$ 394.71	$ 15,605.29	$ 15,733.34
2	501.38	$ 104.04	$ 397.34	$ 15,207.95	$ 15,471.12
3	501.38	$ 101.39	$ 399.99	$ 14,807.96	$ 15,213.27
4	501.38	$ 98.72	$ 402.66	$ 14,405.30	$ 14,959.71
5	501.38	$ 96.04	$ 405.34	$ 13,999.96	$ 14,710.38
6	501.38	$ 93.33	$ 408.05	$ 13,591.91	$ 14,465.21
7	501.38	$ 90.61	$ 410.77	$ 13,181.14	$ 14,224.12
8	501.38	$ 87.87	$ 413.51	$ 12,767.63	$ 13,987.05
9	501.38	$ 85.12	$ 416.26	$ 12,351.37	$ 13,753.94
10	501.38	$ 82.34	$ 419.04	$ 11,932.33	$ 13,524.70
11	501.38	$ 79.55	$ 421.83	$ 11,510.50	$ 13,299.29
12	501.38	$ 76.74	$ 424.64	$ 11,085.86	$ 13,077.64
13	501.38	$ 73.91	$ 427.47	$ 10,658.39	$ 12,859.68
14	501.38	$ 71.06	$ 430.32	$ 10,228.07	$ 12,645.35
15	501.38	$ 68.19	$ 433.19	$ 9,794.88	$ 12,434.59
16	501.38	$ 65.30	$ 436.08	$ 9,358.80	$ 12,227.35
17	501.38	$ 62.39	$ 438.99	$ 8,919.81	$ 12,023.56
18	501.38	$ 59.47	$ 441.91	$ 8,477.90	$ 11,823.17
19	501.38	$ 56.52	$ 444.86	$ 8,033.04	$ 11,626.12
20	501.38	$ 53.55	$ 447.83	$ 7,585.21	$ 11,432.35
21	501.38	$ 50.57	$ 450.81	$ 7,134.40	$ 11,241.81
22	501.38	$ 47.56	$ 453.82	$ 6,680.58	$ 11,054.44
23	501.38	$ 44.54	$ 456.84	$ 6,223.74	$ 10,870.20
24	501.38	$ 41.49	$ 459.89	$ 5,763.85	$ 10,689.03
25	501.38	$ 38.43	$ 462.95	$ 5,300.90	$ 10,510.88
26	501.38	$ 35.34	$ 466.04	$ 4,834.86	$ 10,335.70
27	501.38	$ 32.23	$ 469.15	$ 4,365.71	$ 10,163.44
28	501.38	$ 29.10	$ 472.28	$ 3,893.43	$ 9,994.05
29	501.38	$ 25.96	$ 475.42	$ 3,418.01	$ 9,827.48
30	501.38	$ 22.79	$ 478.59	$ 2,939.42	$ 9,663.69
31	501.38	$ 19.60	$ 481.78	$ 2,457.64	$ 9,502.63
32	501.38	$ 16.38	$ 485.00	$ 1,972.64	$ 9,344.25
33	501.38	$ 13.15	$ 488.23	$ 1,484.41	$ 9,188.51
34	501.38	$ 9.90	$ 491.48	$ 992.93	$ 9,035.37
35	501.38	$ 6.62	$ 494.76	$ 498.17	$ 8,884.78
36	501.38	$ 3.32	$ 498.06	$ 0.11	$ 8,736.70

You are never upside down. For the first 3 months you would owe extra with a $500 deductible. (refers to months 1–10)

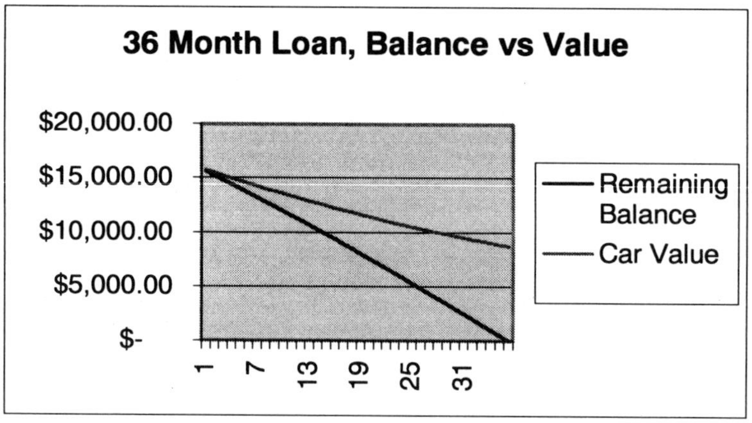

Exploration 5: Rent to Own

1. Payment for $899, 12%, 32 months = 32.97 per month

2. Monthly payment is much less than 4 times the weekly payment.

3. $35.58

4. Interest rate for $29.99 for 130 weeks = 170.875%

Chapter 2

Exploration 6: Multiplying Rabbits

1.

1	2	3	4	5	6	7	8	9	10
150	171	195	222	252	286	324	367	414	466
11	12	13	14	15	16	17	18	19	20
522	584	652	724	801	883	968	1057	1149	1242
21	22	23	24	25	26	27	28	29	30
1336	1429	1521	1610	1696	1778	1855	1927	1993	2054
31	32	33	34	35	36	37	38	39	40
2109	2158	2202	2242	2276	2307	2334	2357	2377	2395
41	42	43	44	45	46	47	48	49	50
2410	2423	2434	2444	2452	2459	2465	2470	2475	2478

3 Graph

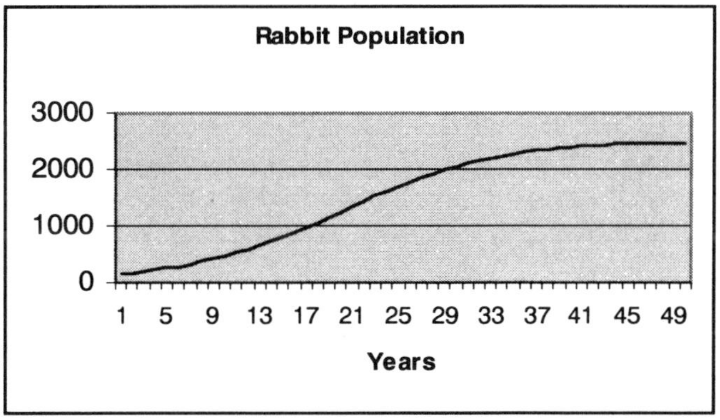

4. 2478

5. 95% of 2500 = 2375 => surpassed in 38th year

Extensions:
Larger r => reaches carrying capacity sooner.

Larger K => stretches the curve both vertically up and horizontally to the right

Larger N => reaches carrying capacity sooner

Exploration 7: The Phone's for You

1.

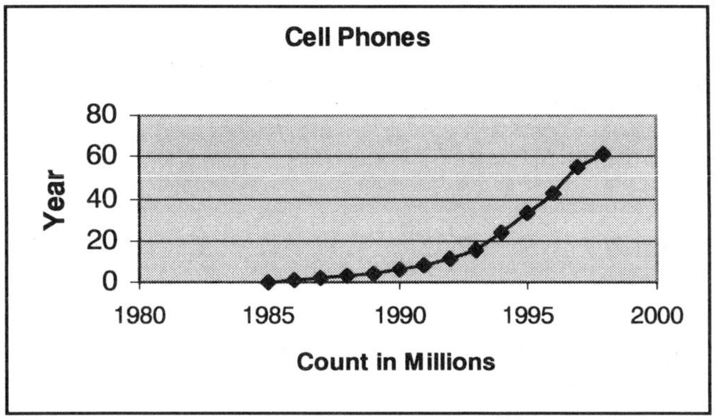

3.

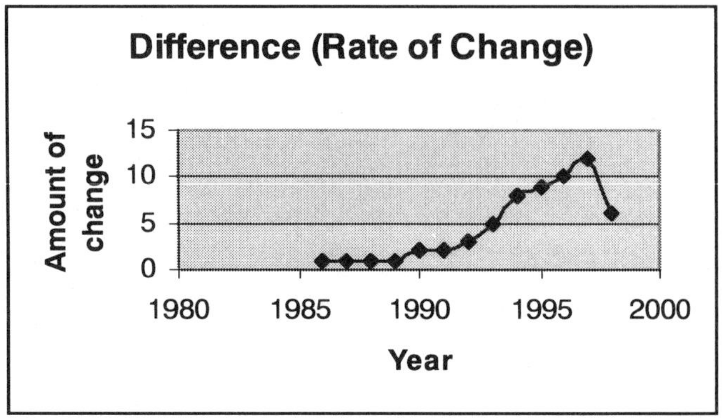

2. Cell_phone(t) = $-.00757t^4 + 0.2449t^3 - 2.1213t^2 + 7.3906t - 6.2722$

(where t = number of years since 1985)

4. Difference_function (t) = $-0.03409t^3 + 0.8362t^2 - 5.1018t + 9.1948$

Exploration 8: Predicting Winners

1.

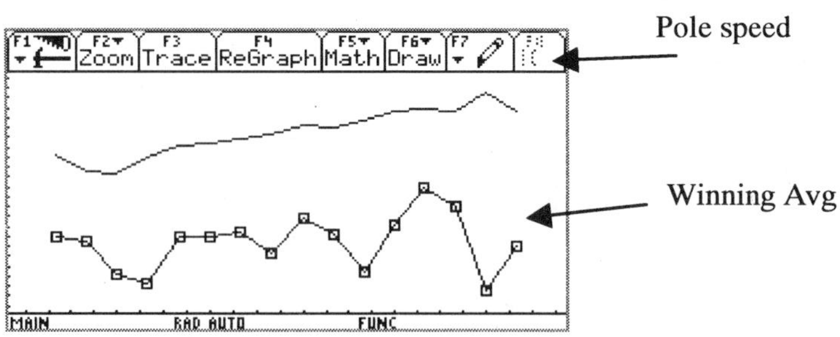

2 and 4.

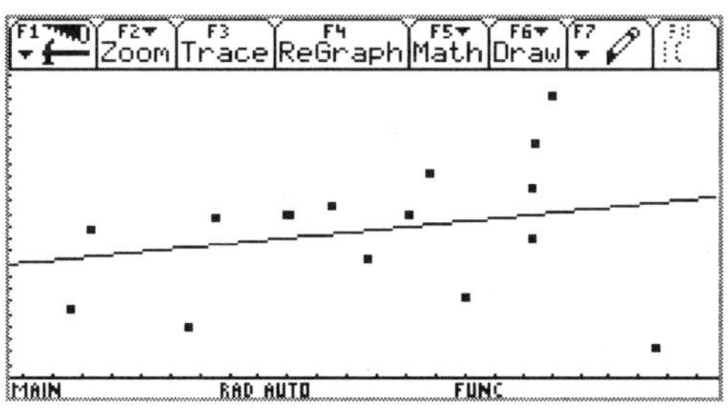

3. Linear Regression Winning_MPH = 0..285178.*x + 98.1736, correlation = .247122

4. 1975 f(193.976) = 153.491 off by 4.278

5. 1983 f(x) = 162.117 ⇒ x = 224.223 off by 19.828

6. 1951 f(x) = 126.244 ⇒ x = 98.431 off by –38.067

Exploration 9: How much faster are they going faster?

1.

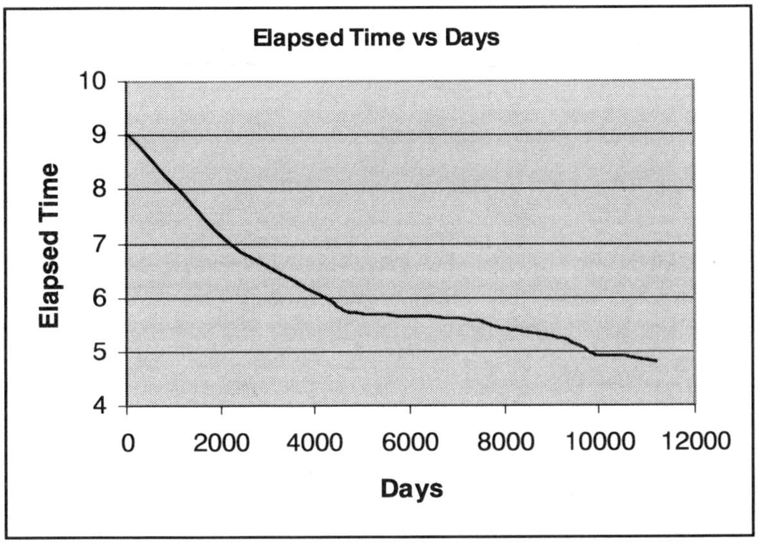

Date	Elapsed time	Change in Days	Change in ET	Rate of change
18-Feb-61	9.000			
20-May-61	8.970	92	-0.030	-3.2609E-04
16-May-64	7.960	1076	-1.010	-9.3866E-04
7-May-67	6.970	1071	-0.990	-9.2437E-04
17-Nov-72	5.970	1990	-1.000	-5.0251E-04
18-Nov-73	5.780	361	-0.190	-5.2632E-04
10-Oct-75	5.698	682	-0.082	-1.2023E-04
18-Oct-81	5.573	2168	-0.063	-2.9059E-05
6-Sep-82	5.454	318	-0.119	-3.7421E-04
1-Oct-83	5.391	385	-0.063	-1.6364E-04
25-Sep-86	5.250	1074	-0.141	-1.3128E-04
4-Apr-87	5.176	189	-0.074	-3.9153E-04
1-Oct-87	5.090	177	-0.086	-4.8588E-04
9-Apr-88	4.990	188	-0.100	-5.3191E-04
9-Oct-88	4.936	180	-0.054	-3.0000E-04
7-Oct-89	4.919	358	-0.017	-4.7486E-05
29-Sep-90	4.881	352	-0.038	-1.0795E-04
22-Mar-92	4.801	533	-0.080	-1.5009E-04

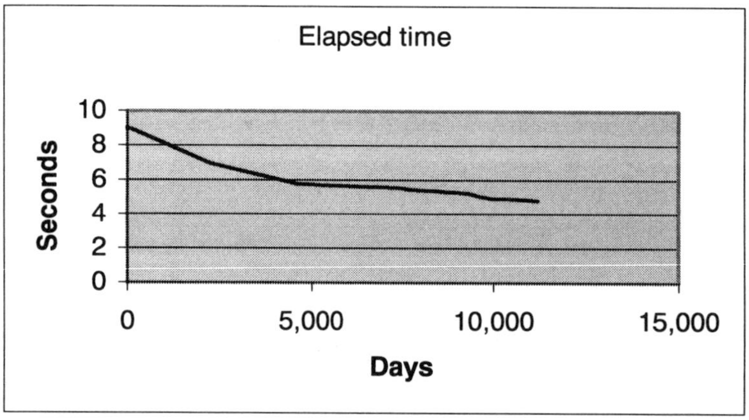

2. Generally concave up, but almost linear, given the size of the domain.

3. $-6.65\times10^{-12}x^3+1.504\times10^{-7}x^2-0.00124x+9.076$ (which is basically linear)

4.

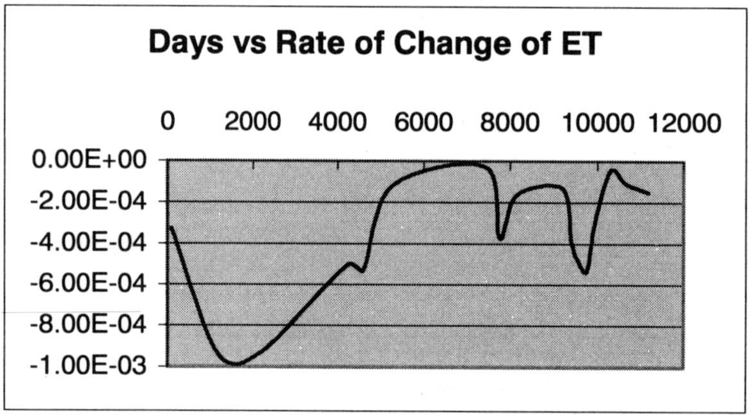

5. 16-May-64, largest decrease in elapsed time. New record set (old record broken) by largest amount.

Exploration 10: What a Drag!!

1.

ET	4.879	4.814	4.746	4.778
Final Speed	292.87	296.15	307.9	312.5
Avg Speed	184.464	186.9547	189.6334	188.3633

ET	5.706	5.692	5.613	5.623	5.625	5.78
Final Speed	241.44	248.34	250	239.23	248.81	249.65
Avg Speed	157.7287	158.1167	160.3421	160.0569	160	155.7093

2.

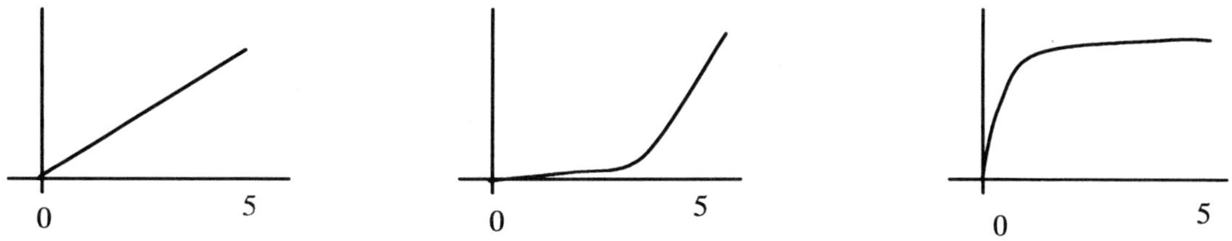

3.

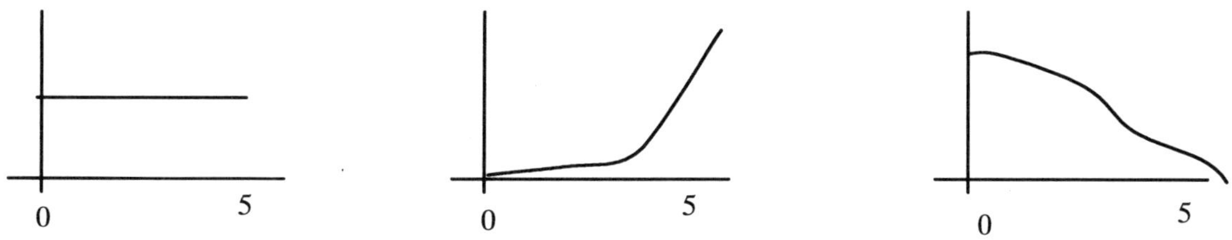

It is possible to have a higher final speed and lose the race – note the middle graphs. The best average speed is always the winner because it deals with elapsed time – the first person across the line and the quarter mile, a fixed distance.

Exploration 11: Swing Batter!!

1. 2. & 3.

Speed (MPH)	75	80	85	90	95	100	
Time Elapsed	0.227274	0.213069	0.200536	0.189395	0.179427	0.170455	25 ft.
	0.272729	0.255683	0.240643	0.227274	0.215312	0.204546	30 ft.
Angular Velocity	10.36719	11.05833	11.74948	12.44063	13.13177	13.82292	25 ft.
	8.639324	9.215279	9.791234	10.36719	10.94314	11.5191	30 ft.
Linear Velocity (feet per second)	51.83594	55.29167	58.7474	62.20313	65.65886	69.11459	25 ft.
	43.19662	46.0764	48.95617	51.83594	54.71572	57.59549	30 ft.

4a. Swing as usual, it will be too soon … miss or hit foul left

4b. Swing later or swing slower (or both)

Extension

Speed	75	80	85	90	95	100	
Pitching Time	0.578541	0.542382	0.510477	0.482117	0.456743	0.433905	
Catcher Ready Time	0.5	0.5	0.5	0.5	0.5	0.5	
Catcher Throw Time	1.157082	1.157082	1.157082	1.157082	1.157082	1.157082	(Assume Catcher 75 mph throw)
Total	2.235623	2.199464	2.167559	2.139199	2.113825	2.090987	

Runner steals second safely in 2.903 seconds, only if a bad throw to defensive man on 2nd.

Exploration 12: But I Never Inhaled

2. Pulse rate for lettuce cigarette appears possibly linear.
3.

minutes	Pulse rate lettuce	Temp lettuce	Pulse rate Tobacco	Temp Tobacco
0	64	33.4	66	32.5
4	65	32	74	31
8	66	31	81	30
12	64	30.8	81	29.1
16	65	30.5	81	28.5
20	66	30.6	81	28
24	66	30.7	80	27.9
28	64	30.9	80	27.75
32	63	31	76	27.5

4. $P(t) = -.20833t + 65.111$ however, it misses the points. They are alternatively above and below the modeling line. It can only show general tendency.

Good models for the skin temperature turn out to be quadratic.

$Tlettuce(t) = .0065t^2 - .2651t + 33.095$

$Ttobacco(t) = .00591t^2 - .33511t\ 32.358$

Exploration 13: How Do You Know How Low You Can Go?

$$altitude(\theta) = \frac{75}{2}\tan\theta$$

Angle	15	20	25	30	35	40	45	50	55	60	65	70	75	80
Altitude	10.05	13.65	17.49	21.65	26.26	31.47	37.5	44.69	53.56	64.95	80.42	103	140	212.7
Difference	3.601	3.838	4.164	4.607	5.208	6.034	7.191	8.865	11.4	15.47	22.61	36.92	72.72	

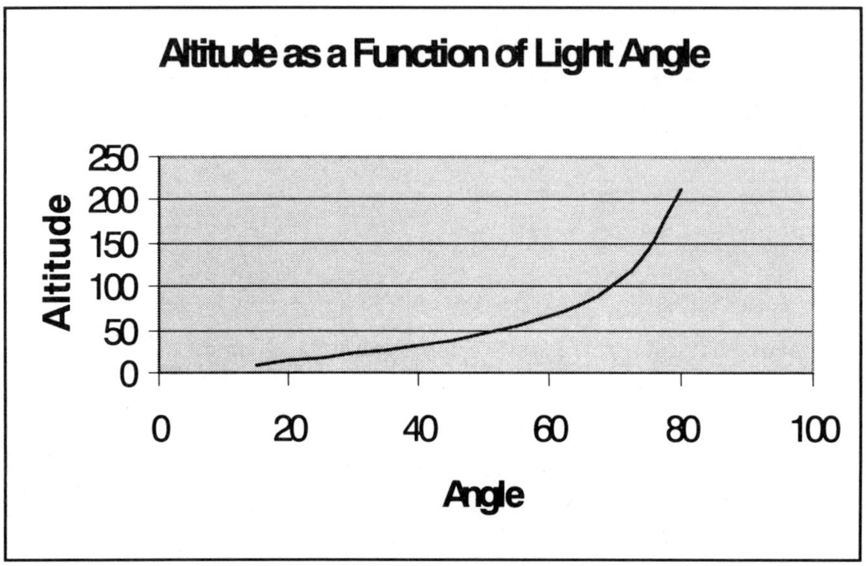

6. Units: feet per degree

Extension:

Angle	15	20	25	30	35	40	45	50	55	60	65	70	75	80
Altitude	5.80	7.88	10.10	12.5	15.16	18.17	21.65	25.80	30.92	37.7	46.43	59.49	80.80	122.79
Difference	2.08	2.21	2.40	2.66	3.01	3.48	4.15	5.12	6.58	8.93	13.06	21.32	41.99	

Exploration 14: What's Hidden in the π?

1.

Digit	1	4	1	5	9	2	6	5	3	5	8	9	7
greater - less	12-0	9-3	11-0	6-2	0-7	7-0	3-3	3-1	4-0	3-0	1-1	1-0	0-0
Diff.	12	6	11	3	-7	7	0	2	4	3	0	1	0

Sum of Diff, S= 42   n = 14

$z = 2.29929$

Chapter 3

Exploration 15: Uncle Sam Wants to Count YOU!

Ratios:

1840	17.1		2167		833	
1850	23.2	1.36	3231	1.49	1423	1.71
1860	31.4	1.35	4417	1.37	1969	1.38
1870	38.6	1.23	6530	1.48	3421	1.74
1880	50.2	1.30	31382	4.81	5790	1.69
1890	63	1.25	46804	1.49	11547	1.99
1900	76	1.21	52871	1.13	11854	1.03
1910	92.2	1.21	70286	1.33	15968	1.35
1920	106	1.15	87234	1.24	25117	1.57
1930	123.2	1.16	87756	1.01	40156	1.60
1940	132.2	1.07	123069	1.40	67527	1.68
1950	151.3	1.14	142962	1.16	91462	1.35
1960	179.3	1.19	159321	1.11	127934	1.40
1970	203.2	1.13	168406	1.06	247653	1.94
1980	228.5	1.12	458523	2.72	1136000	4.59
1990	248.7	1.09	510200	1.11	2600000	2.29

1. Graphs:

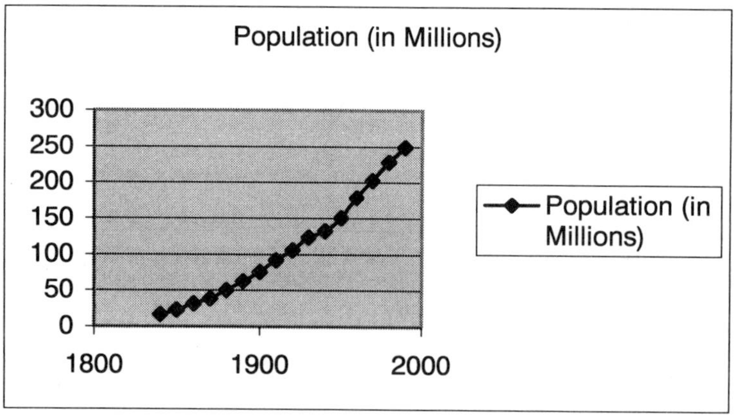

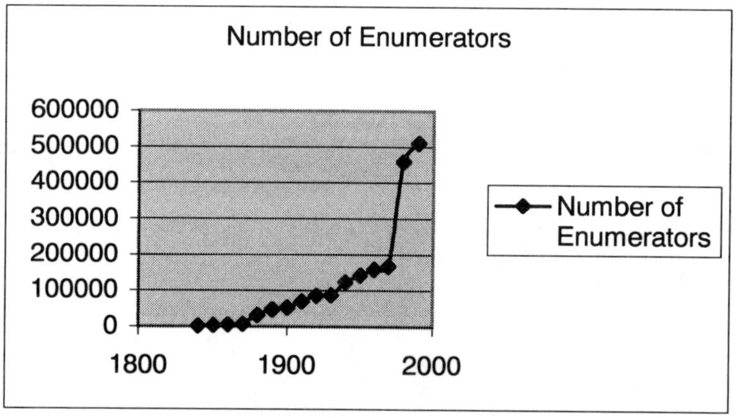

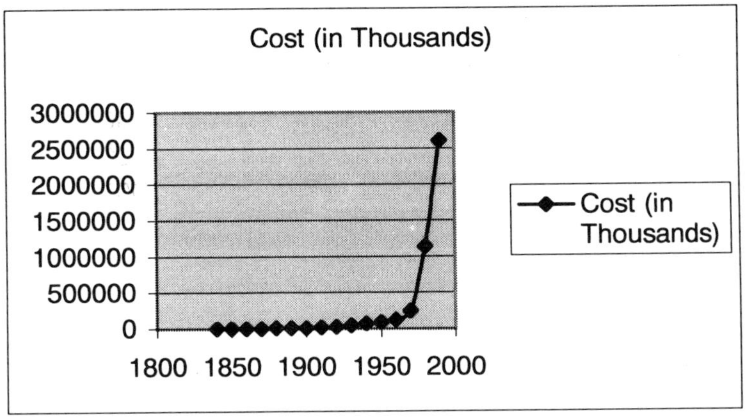

5. Regression Equations:
Where x is the census year.

Population:

$y = (3.796 \times 10^{-13})(1.0174)^x$

Enumerators:

$y = (1.867 \times 10^{-24})(1.0348)^x$

Cost:

$y = (3.988 \times 10^{-36})(1.049)^x$

Exploration 16: Spooky October Stock Market

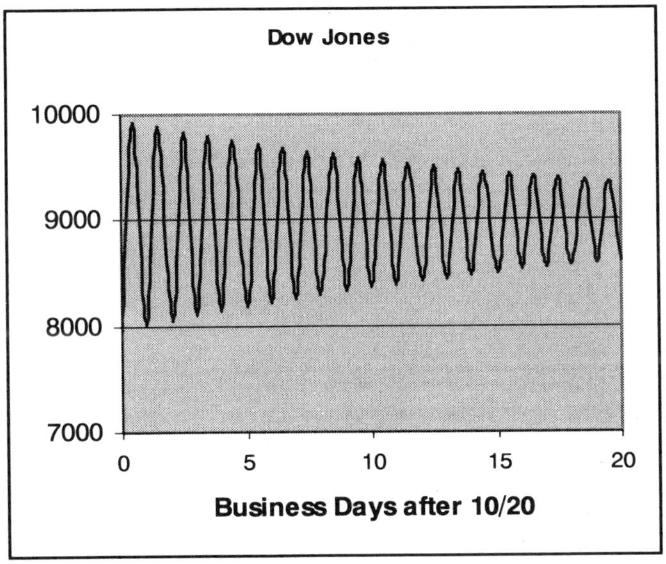

D(3) = 8097.94

D(13.05) = 8479

Day = t	D(t)	Difference
19	8596.709	21.37864
20	8617.121	20.41158
21	8636.612	19.49169
35	8834.476	10.45958
36	8844.501	10.02506
37	8854.113	9.611722

C3E17: Come to Disney World with Bill Gates!

The formula is New emails (d) = $10 * 1.65^{(d-1)}$, where d is the number of the day.

Day	New emails	Total Emails
1	10	10
2	17	27
3	27	44
4	45	72
5	74	119
6	122	196
7	202	324
8	333	535
9	549	882
10	906	1,456
11	1,496	2,402
12	2,468	3,964
13	4,072	6,540
14	6,719	10,791
15	11,086	17,805
16	18,292	29,378
17	30,182	48,474
18	49,800	79,981
19	82,170	131,969
20	135,580	217,749
21	223,707	359,286
22	369,116	592,823
23	609,041	978,157
24	**1,004,918**	**1,613,960**
25	1,658,115	2,663,033
26	2,735,890	4,394,005
27	4,514,218	7,250,108
28	7,448,460	11,962,678
29	12,289,959	19,738,419
30	20,278,433	32,568,392
31	33,459,414	53,737,846
32	55,208,033	88,667,446
33	91,093,254	146,301,286
34	150,303,869	241,397,122
35	248,001,383	398,305,252
36	409,202,282	657,203,666
37	**675,183,766**	**1,084,386,049**
38	1,114,053,214	1,789,236,980
39	1,838,187,803	2,952,241,017
40	3,033,009,875	4,871,197,678
41	5,004,466,294	8,037,476,169
42	8,257,369,385	13,261,835,680
43	13,624,659,486	21,882,028,871
44	22,480,688,152	36,105,347,637
45	37,093,135,450	59,573,823,602

7. Day 25
8. Day 37

Exploration 18: Don't Forget the Sunscreen

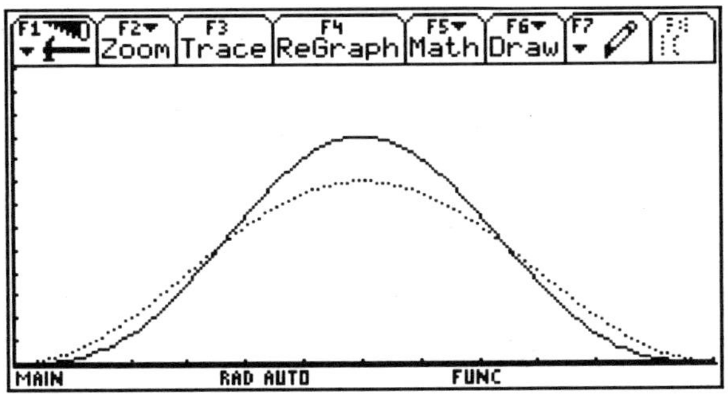

The dotted line is the function for the intensity on a cloudy day.

Graph of $I_{cloudy} - I_{sunny}$

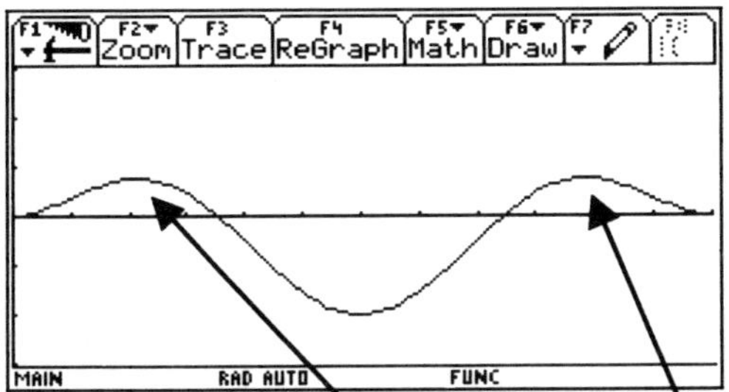

About 2.15 hours after sunrise or before sunset, cloudy has the most intensity greater than the sunny day, or 32.231º above the horizon.

Slope of Tangent lines for hours 0 - 6

Time	0	1	2	3	4	5	6
Sunny Day	0	.50819	1.7004	2.7768	2.9452	1.8966	0
Cloudy Day	0	1.0472	1.8138	2.0944	1.8138	1.0472	0

Exploration 19: I Object, Your Honor!

1.

Linear Increase			
Number Released	Number of Crimes	Number Imprisoned	Number of Crimes
0	650	0	650
10	662	10	639
20	673	20	627
30	685	30	616
40	696	40	604
50	708	50	593
60	719	60	581
70	731	70	570
80	742	80	558
90	754	90	547
100	765	100	535
110	777	110	524
120	788	120	512
130	800	130	501
140	811	140	489
150	823	150	478
60	834	160	466
170	846	170	455
180	857	180	443
190	869	190	432
200	880	200	420

3.

Exponential Increase			
Number Released	Number of Crimes	Number Imprisoned	Number of Crimes
0	650	0	650
10	660	10	640
20	670	20	631
30	680	30	621
40	690	40	612
50	701	50	603
60	711	60	594
70	722	70	585
80	733	80	576
90	744	90	568
100	755	100	559
110	767	110	551
120	778	120	543
130	790	130	535
140	802	140	527
150	814	150	519
160	826	160	511
170	839	170	504
180	851	180	496
190	864	190	489
200	877	200	481

Functions:
2a $y = 650 + 1.15x$
3a $y = 650\ (1.0015)^x$
2b $y = 650\ (.9985)^x$
3b $y = 650 - 1.15x$

Graph: from top to bottom in the order above

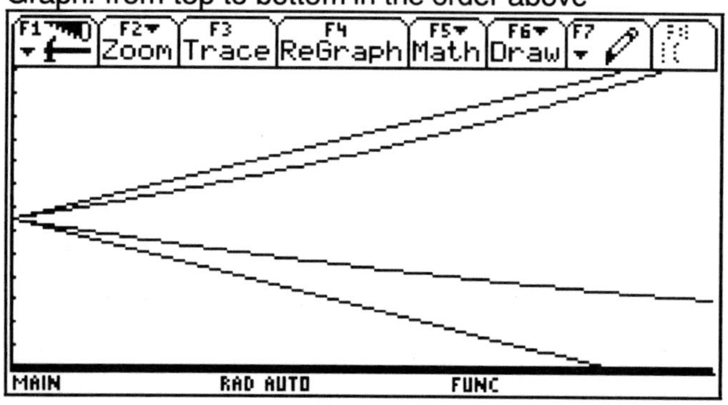

Double when x = 545 released for linear, 462 for exponential
Zero when x = 565 additional added for linear. Would never reach zero for exponential.

Exploration 20; Two Bolas Make a Parabola

1. Standard Window

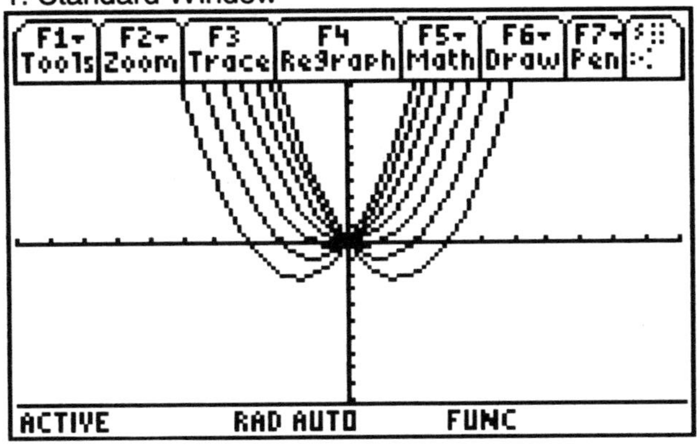

2. They all have the y-intercept (0,0).

3. They have different vertex values.

4.

F1 | F2 Plot Setup | F3 Cell | F4 Header | F5 Calc | F6 Util | F7 Stat

DATA

	c1	c2	c3	c4	c5
1	-3	1.5	.75		
2	-2	1.	2.		
3	-1	.5	2.75		
4	0	0.	3.		
5	1	-.5	2.75		
6	2	-1.	2.		
7	3	-1.5	.75		

c3=c2^2+c1*c2+3.

MAIN RAD AUTO FUNC

5. Window is [-5,5 by [-5,5]

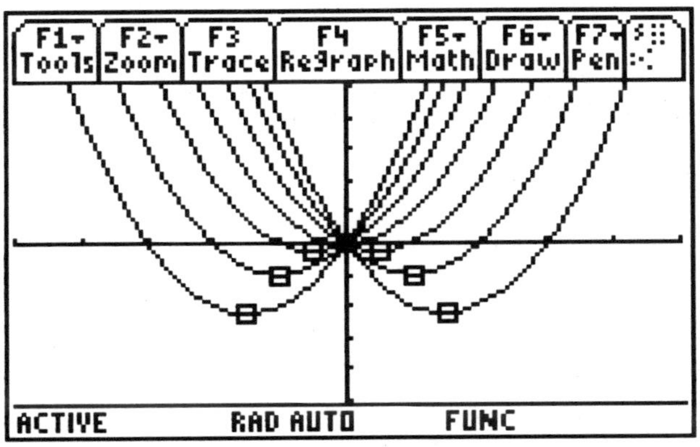

6a. & b. The equation is $f(x) = -x^2 + 3$

Extension 1: *c* moves each graph up or down.

Extension 2: For $2x^2 + bx - 4$

The equation for the vertices $f(x) = -2x^2 - 4$

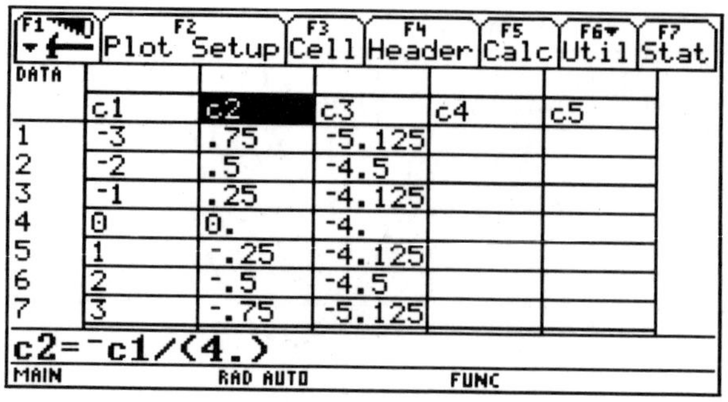

DATA	c1	c2	c3	c4	c5
1	-3	.75	-5.125		
2	-2	.5	-4.5		
3	-1	.25	-4.125		
4	0	0.	-4.		
5	1	-.25	-4.125		
6	2	-.5	-4.5		
7	3	-.75	-5.125		

c2=-c1/(4.)

In general the equation for the vertices is $f(x) = -ax^2 + c$

Exploration 21; Follow the Bouncing Ball

1a. $b(h) = \text{int}\left(\dfrac{-\ln(8*h)}{\ln 0.75}\right)$

1b.

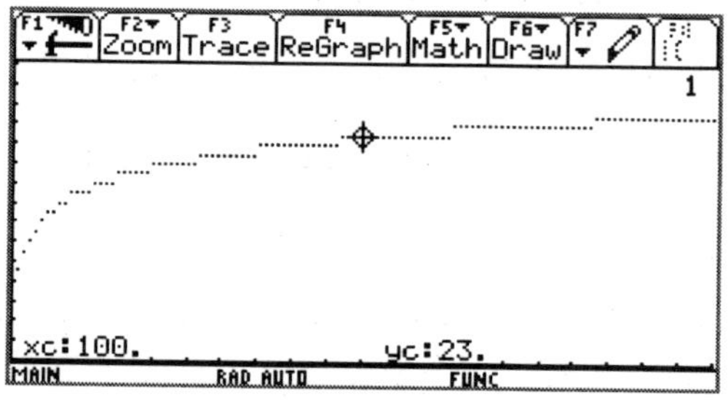

Height after each bounce:

2a. $h(b) = 6*(0.75)^b$

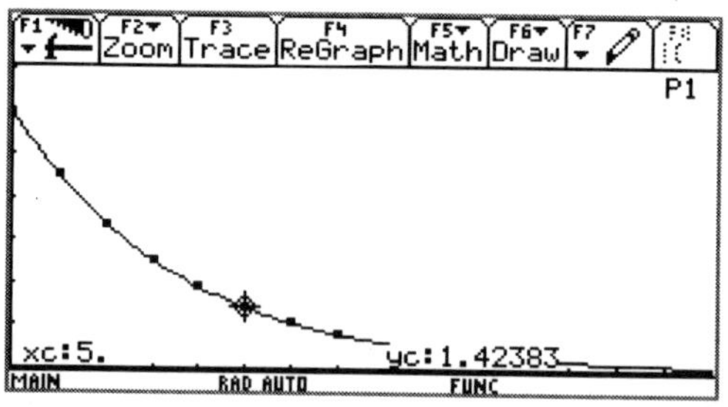

2b.

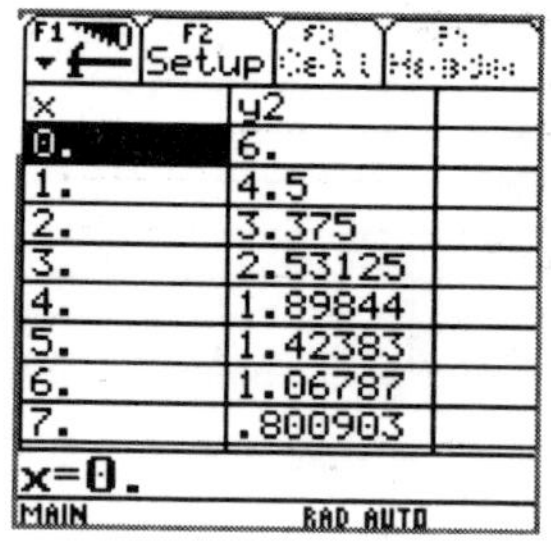

x	y2
0.	6.
1.	4.5
2.	3.375
3.	2.53125
4.	1.89844
5.	1.42383
6.	1.06787
7.	.800903

x=0.

2f. Bounces for highest bounce = ½ inch

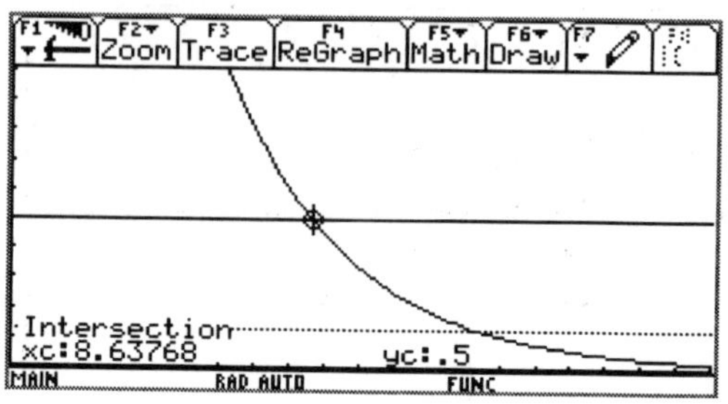

Bounces for highest bounce = 1/8 inch

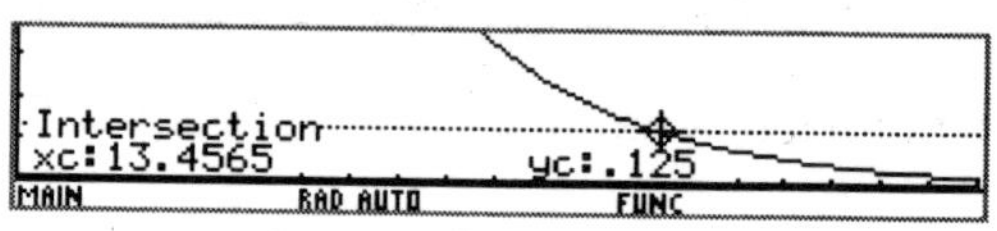

h(b) < 0.01 when b > 22.24 or 23 bounces

Exploration 22: S.pin T.ires O.n P.avement

1. Note: these are the actual values for a 20 second delay used to make the graph, not estimates from the graph. Your estimates should be close to these values.

2.

Approach Speed	10	15	20	25	30	35	40	45	50	55	60
Cost	0.9693	1.0451	1.1673	1.3361	1.5513	1.8131	2.1213	2.4761	2.8773	3.3251	3.8193

Cost (x) = 0.00093*x^2-0.0081*x+0.9573 where x is the approach speed

2.

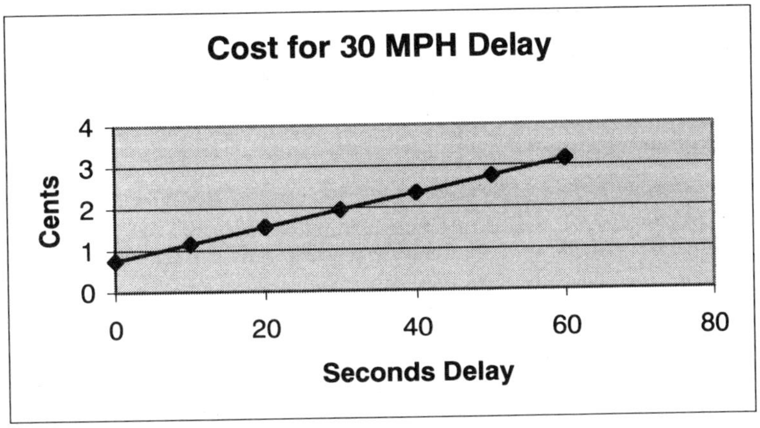

3. Cost for 30 mph approach:

Delay	No Delay	10 sec	20 sec	30 sec	40 sec	50 sec	60 sec
Cost	0.7513	1.1513	1.5513	1.9513	2.3513	2.7513	3.1513

Note the difference between the above costs is a constant 0.4 cents (for an increase of 10 seconds)

4.

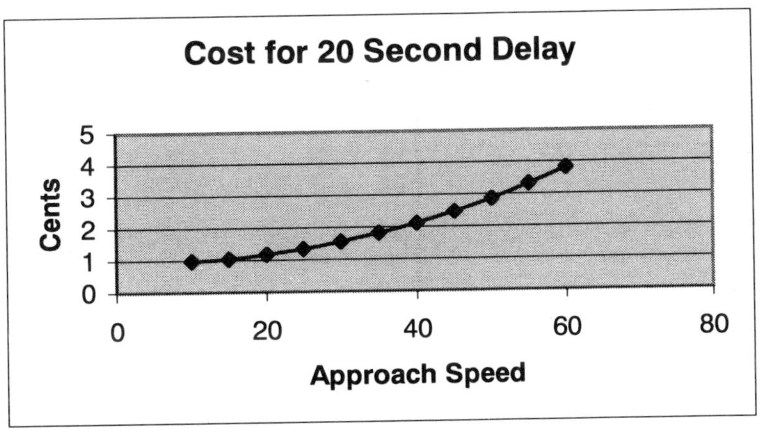

7. Questions
Approach speed is 40 and delay time is 15 seconds = 1.9213
Approach speed is 60 and delay time is 30 seconds = 4.2193
When the cost is 2.5 cents and the delay time was 30 seconds, the approach speed = 39.67 mph
When the cost is 1.7 cents and the approach speed is 15 mph., the delay time w=36.37 se

Exploration 23; Batteries Included

Time	Minutes Charge	Minutes Usage	Percent Charged	
8:30			100.00	
8:35		5	84.55	
9:14	39		92.06	
9:27		13	59.51	
10:02	35		66.27	
10:15		13	42.84	
12:18	123		64.65	
12:35		17	36.54	
14:43	128		59.13	
14:59		16	34.56	38.23
15:30	31		40.58	44.24
15:36		6	33.18	36.17
16:19	43		41.42	44.41
16:28		9	30.62	32.83
17:12	44		39.04	41.26
17:19		7	30.87	32.62
8:30	911		100.00	100.00

2. Going with total usage during the day (from 8:30 to 17:19), with 83 minutes of usage, it discharges the battery down to 6.17 % and 443 minutes of fuel cell charging, it runs it back from 6.17% charge, increasing it by 57.77% to 64.94% -- not the same as the results shown in the table above.

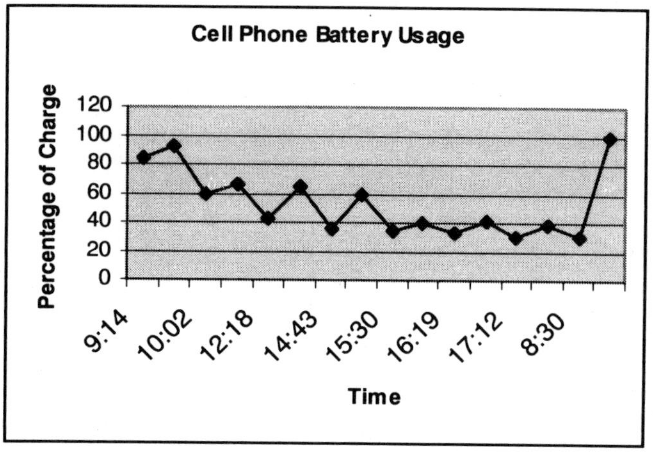

Exploration 24: Chili Today, Hot Tamale

Wind Speed \ Temperature	35	30	25	20	15	10	5	0	-5
5	32	27	22	16	11	6	1	-5	-10
10	22	16	10	4	-2	-9	-15	-21	-27
15	16	9	2	-4	-11	-18	-25	-31	-38
20	11	4	-3	-10	-17	-24	-31	-39	-46
25	8	1	-7	-14	-22	-29	-36	-44	-51
30	6	-2	-10	-17	-25	-32	-40	-48	-55
35	4	-4	-12	-19	-27	-35	-43	-51	-58

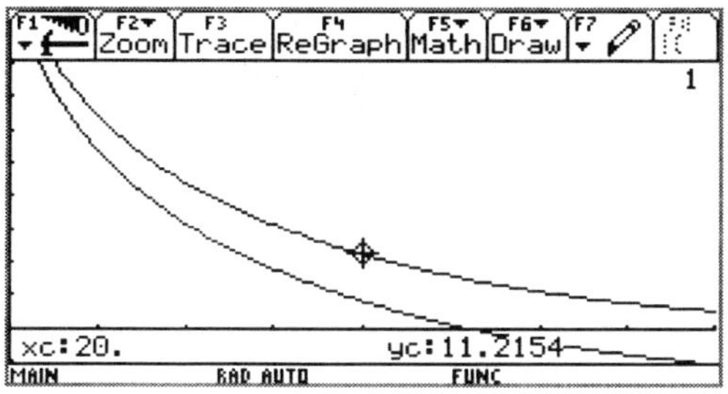

Temp constant, vary wind speed

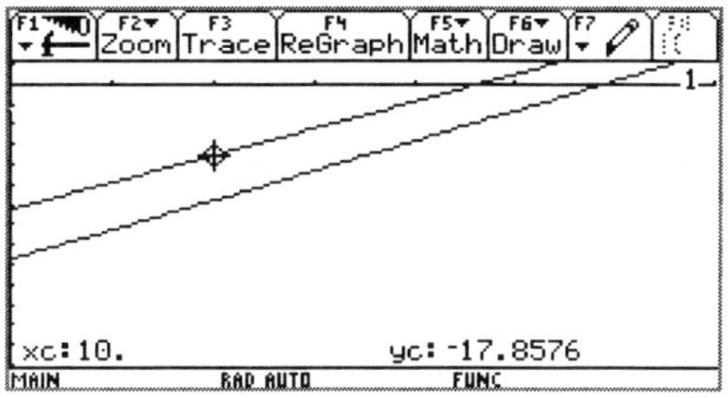

Wind speed constant, vary temp

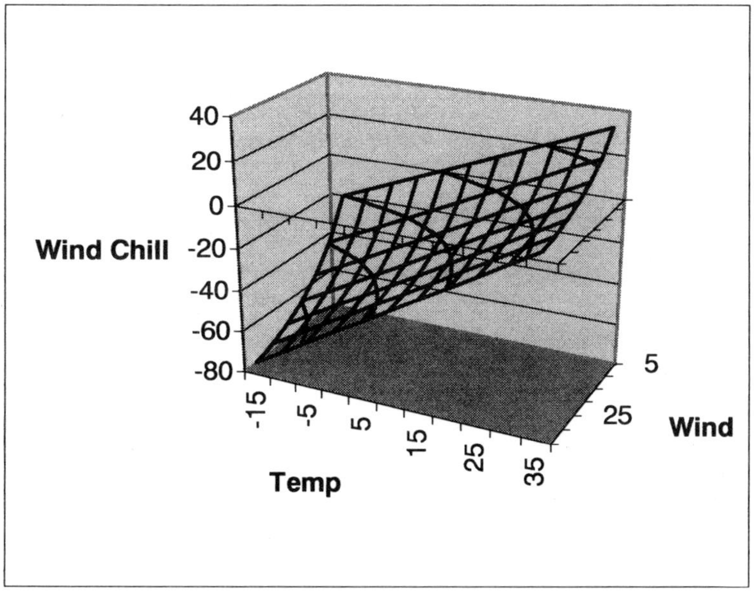

Chapter 4

Exploration 25: Two Hands Are Better Than One

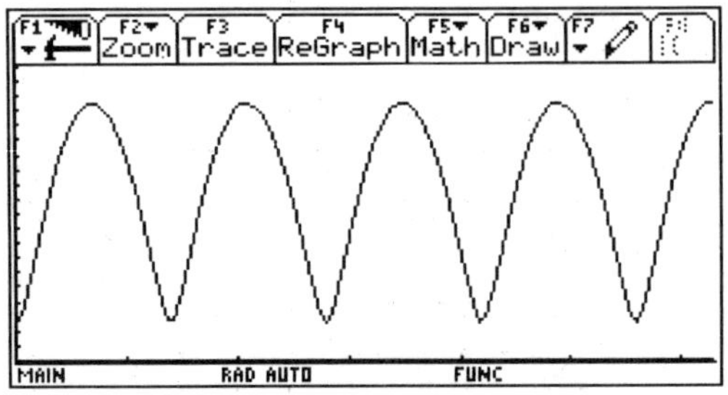

For the second hand:
xt1(t)=1 * sin(t)
yt1(t)=1*cos(t)

For the 1/10 second hand:
xt2(t)=0.75*sin(10t)
yt2(t)=0.75*cos(10t)

dist(x)=sqrt((x1t(x)-x2t(x))^2+(y1t(x)-y2t(x))^2)

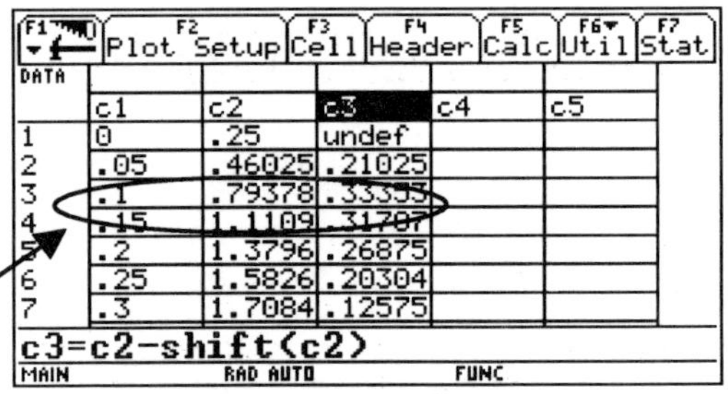

Distance graph for 0<x<π (30 sec)

Minimize 10 times in 60 sec
Maximize 9 times in 60 sec

Minimize 100 times in 10 minutes (600 seconds)

Greatest Difference (steepest point of graph above)

Graph of the difference maximum
minimum

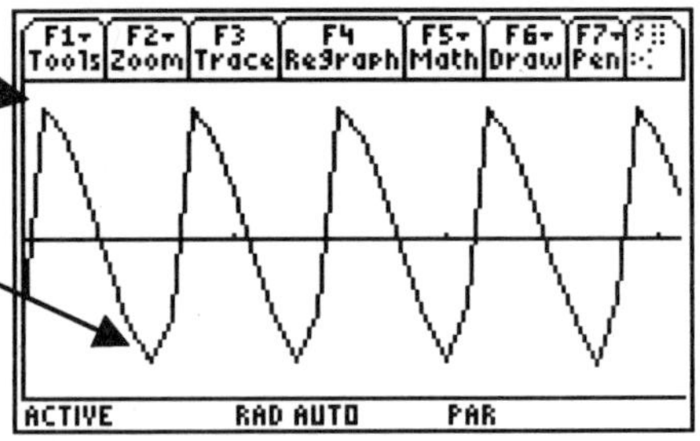

Exploration 26: Incoming Pumpkin

1a. 100 cos 10° or about 9.98 horizontal feet per sec
1b. h(t)= 100*t* cos(θ)
1c. v(t)=100*t*sin(θ)
1d. v(t)=100*t* sin(θ) – 16t^2+4
1e.

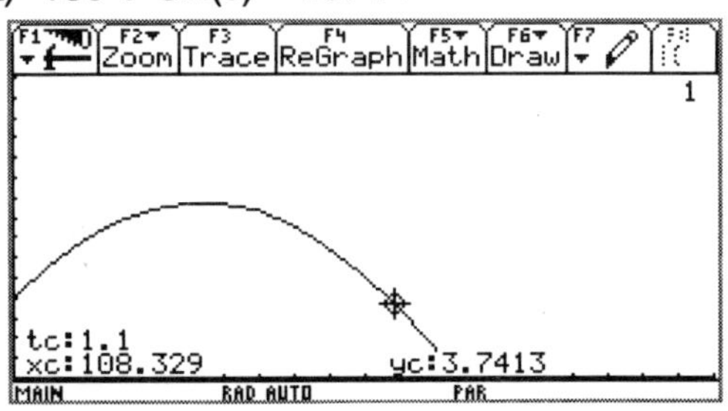

2. 1.28 seconds

3.

θ	0	10	20	30	40	50	60	70	80	90
Distance	50	126	211	277	312	311	272.9	202	107.6	0

Highest vertical distance is not farthest horizontal dist.

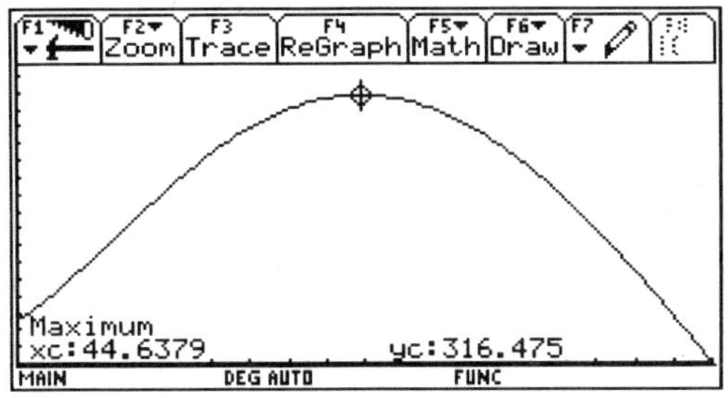

θ for max dist

4. The heavy graph is for 44.64 degrees, the maximum

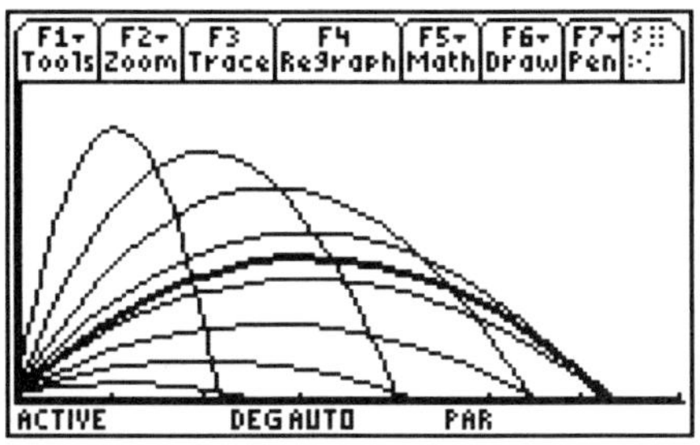

5. Dist as a function of θ

$$dist(\theta) = \frac{\sqrt{625 * \sin^2 \theta + 16} + 25 * \sin\theta}{8} * 100 * \cos\theta$$

Exploration 27: Have You Hugged Your Crash Dummy Today?

1. For the center point of the air bag:
xt1(t) = 20 * t * cos (20)
yt1(t) = 20 * t * sin (20)

For the front of the chest of the driver:
xt2(t) = -95.334t + 1.333
yt2(t) = 0

3. xt1(t) = xt2(t) when t = .011683 (at .219565 ft from the original center point of the steering wheel)

4. no

5. To animate when the chest and air bag come together:

$$xt1(t) = \begin{cases} 20*t*\cos(20) & \text{for } t \le .05 \\ -3.33*t*\cos(20) & \text{for } 0.5 < t \le 0.3 \\ 0 & \text{for } t > 0.3 \end{cases}$$

$$xt2(t) = \begin{cases} -95.334*t + 1.333 & \text{for } t \le .011683 \\ xt1(t) & \text{for } t > 0.11683 \end{cases}$$

Exploration 28: Round and Round

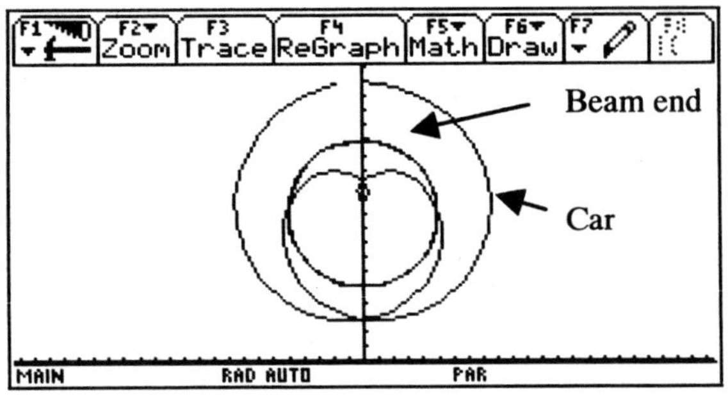

Equations for each end of the beam:

xt1(t) = 25*sin(t) yt1(t) = 25*cos(t)+50
xt2(t) = - xt1(t) yt2(t) = -25*cos(t) + 50

Equations for each of the cars:

xt3(t) = xt1(t) + 20*sin(1.5*t) yt3(t) = yt1(t) + 20*cos(1.5*t)
xt4(t) = xt2(t) – 20*sin(1.5*t) yt4(t) = yt2(t) + 20*cos(1.5*t)

Graph for two revolutions of the beam, 3 revolutions of the car around the end of the beam.

0	0	95		
1.5707	39.14418	35.86232	39.14418	-59.1377
2.0707	22.65142	18.02911	-16.4928	-17.8332
2.5707	0.405433	13.85555	-22.246	-4.17356
3.0707	-18.1162	22.94002	-18.5216	9.084474
3.5707	-26.3996	39.26912	-8.28337	16.3291
4.0707	-23.5514	54.72332	2.848204	15.4542
4.5707	-13.9086	63.27665	9.642718	8.553331
5.0707	-4.02382	63.67531	9.88482	0.398664
5.5707	1.188868	59.29369	5.212689	-4.38162
6.0707	0.994928	55.44505	-0.19394	-3.84865
6.5707	-1.27129	55.8051	-2.26622	0.360046
7.0707	-0.78675	60.04525	0.484538	4.240152
7.5707	5.289001	64.04209	6.075753	3.996841
8.0707	15.5305	62.54255	10.2415	-1.49953
8.5707	24.56204	52.74386	9.031537	-9.79869
9.0707	25.91298	36.68077	1.350941	-16.0631
9.5707	15.88769	20.9229	-10.0253	-15.7579
10.0707	-3.72396	13.5513	-19.6117	-7.37161
10.5707	-25.7282	19.91374	-22.0042	6.362443
11.0707	-40.5722	39.41423	-14.844	19.50049
11.5707	-40.9184	65.14278	-0.34618	25.72855
12.0707	-25.428	86.71325	15.49039	21.57047
12.5707	0.238115	94.99934	25.66611	8.286095
13.0707	25.80803	86.43241	25.56992	-8.56694
13.5707	41.05372	64.70151	15.24569	-21.7309

Increments of $\frac{\pi}{10}$	X position of car	Y position of Car	X difference	Y difference
0	0.0000	95.0000		
0.3141593	**16.8052**	**91.5965**	**16.8052**	-3.4035
0.6283185	30.8750	81.9811	14.0697	-9.6154
0.9424778	39.9792	67.8233	9.1042	-14.1578
1.2566371	**42.7975**	**51.5451**	2.8184	**-16.2782**
1.5707963	39.1421	35.8579	-3.6554	-15.6872
1.8849556	29.9568	23.2534	-9.1854	-12.6044
2.1991149	17.0967	15.5516	-12.8600	-7.7018
2.5132741	2.9389	13.5942	-14.1578	-1.9574
2.8274334	-10.0947	17.1438	-13.0336	3.5495
3.1415927	-20.0000	25.0000	-9.9053	7.8562
3.4557519	-25.5456	35.3034	-5.5456	10.3034
3.7699112	-26.4503	45.9549	-0.9048	10.6515
4.0840704	-23.3541	55.0591	3.0962	9.1042
4.3982297	-17.5961	61.2957	5.7580	6.2366
4.712389	-10.8579	64.1421	6.7382	2.8464
5.0265482	-4.7553	63.9058	6.1026	-0.2364
5.3407075	-0.4717	61.5659	4.2836	-2.3398
5.6548668	1.4857	58.4697	1.9574	-3.0962
5.969026	1.3544	55.9563	-0.1313	-2.5134
6.2831853	0.0000	55.0000	-1.3544	-0.9563
	Max difference		16.8052	-16.2782

Extension
a. 16.8 horizontal feet in $\pi/10$ radians rotation of beam
b. 16.28 vertical feet in $\pi/10$ radians rotation of beam
c. $\pi/10$ radians revolution takes 1.5 sec
d. 11.2 ft/sec horizontal, 10.9 ft/sec vertical
e. Beam at 0 radians for max horizontal acceleration and at 0.4 radians for max vertical acceleration

Exploration 29: How Do You Get There From Here?

2. xt1(t) = 0 yt1(t) = 30t + 10 * cos(15)

3a. Horizontal component of velocity = -150*sin(15)

3b. Vertical component = 150*cos(15)

3c. xt2(t) = 10*sin(15) – 150*sin(15)*t yt2(t) = 150*cos(15)*t

4. xt2(t) = 10*sin(15) – 150*sin(15)*t yt2(t) = 150*t*cos(15)+30*t

5. yt1(t) = yt2(t) at t = 1/15 hours (4 minutes) and this is at 11.6593 nautical miles from the origin.

6. The pilot headed 12.52º to left of the carrier's heading

In general, for distance *d*, plane velocity *p*, ship velocity *s,* they meet at t = d/p.

The y distance from the origin is $y = s * \frac{d}{p} + d * \cos(15)$.

The angle to the left of the ship's heading will be $\arctan\left(\frac{d * \sin(15)}{s * \frac{d}{p} + d * \cos(15)}\right)$

Exploration 30: Log Rhythms, or Half a Log is Better than None

2.

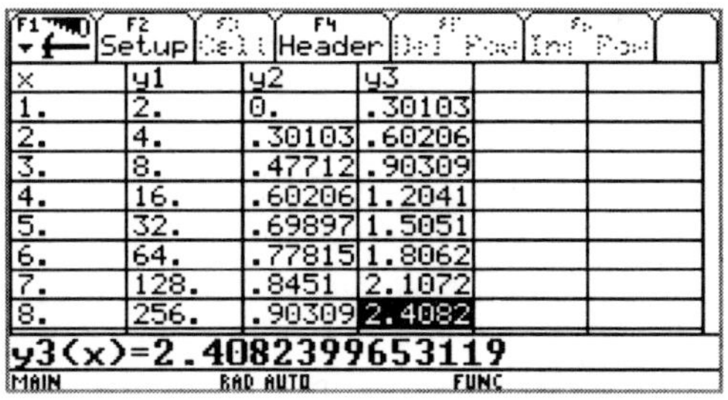

x	y1	y2	y3
1.	2.	0.	.30103
2.	4.	.30103	.60206
3.	8.	.47712	.90309
4.	16.	.60206	1.2041
5.	32.	.69897	1.5051
6.	64.	.77815	1.8062
7.	128.	.8451	2.1072
8.	256.	.90309	2.4082

y3(x)=2.4082399653119

2a.

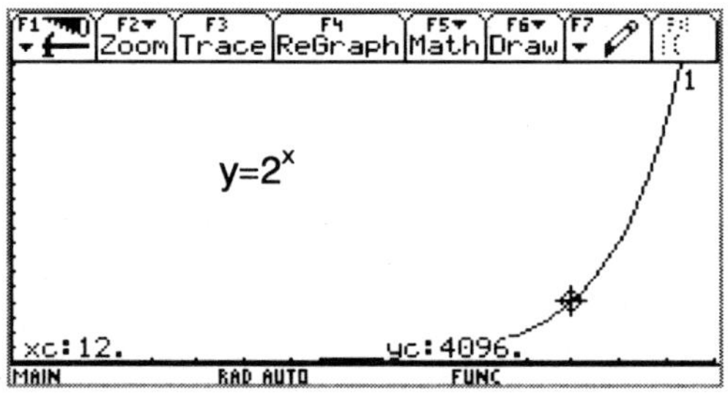

2b.

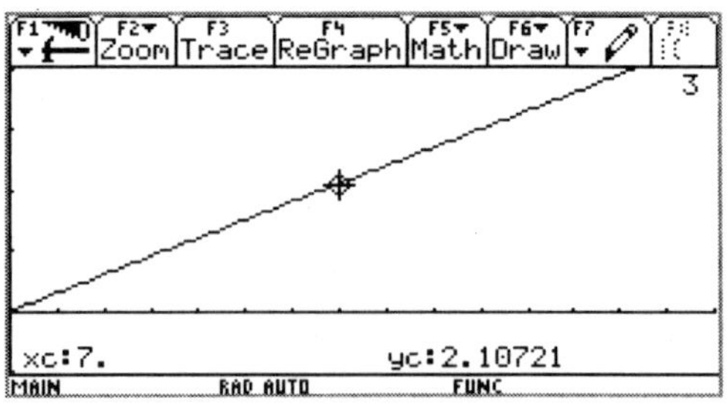

2d. slope = $\log_{10} 2 \approx 0.301$

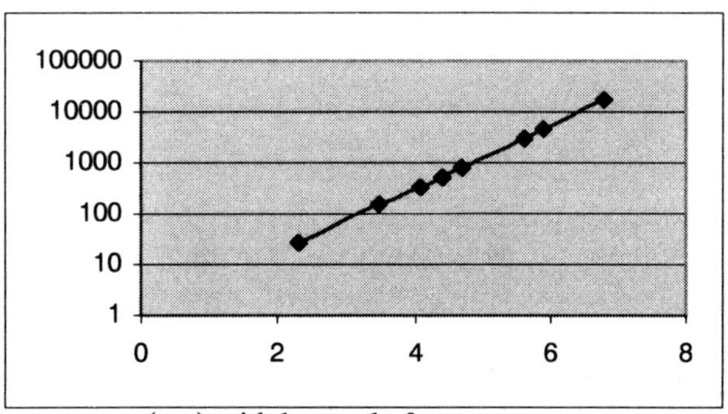

(p,q) with log scale for q

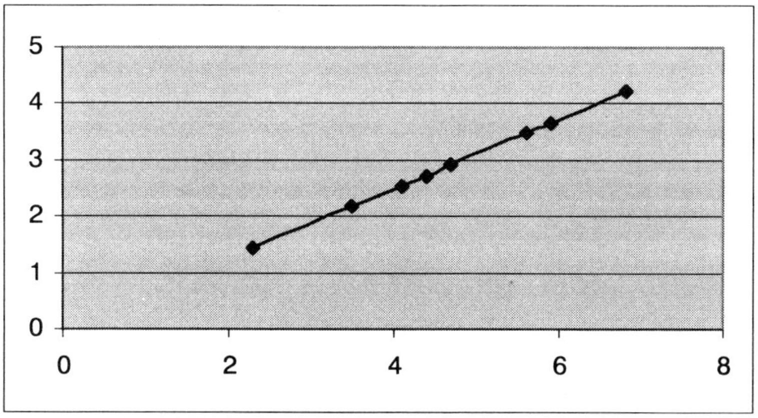

(p,log(q)) with linear scale

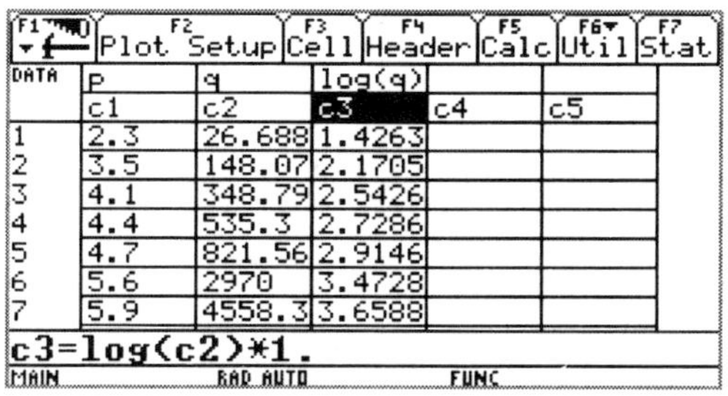

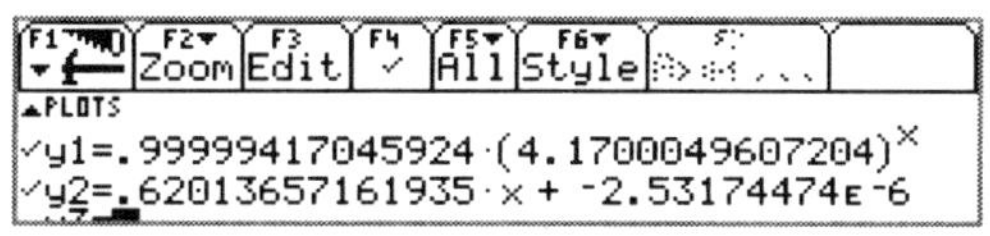

y1 = exponential regression for c1, c2;
y2 = linear regression for c1, c3
Note $\log(4.17) \approx 0.62014)$

With a little rounding: $y1 = 4.17^x$ and $y2 = (\log(4.17))x$. The "underlying function" is $y2 = (\log(4.17))x$

Exploration 31: Log Rhythms, or Why Settle for Half a Log

1a.

x	x^2	$\log x$	$\log(x^2)$
1	1	0.0000	0.0000
2	4	0.3010	0.6021
3	9	0.4771	0.9542
4	16	0.6021	1.2041
5	25	0.6990	1.3979
6	36	0.7782	1.5563
7	49	0.8451	1.6902
8	64	0.9031	1.8062
9	81	0.9542	1.9085
10	100	1.0000	2.0000

1b

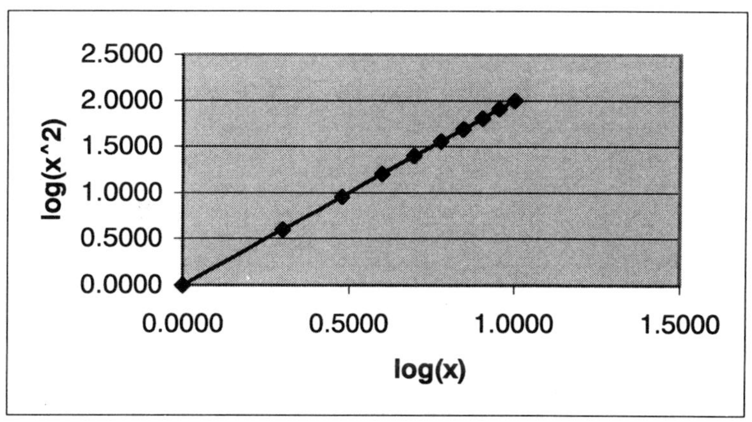

2a.

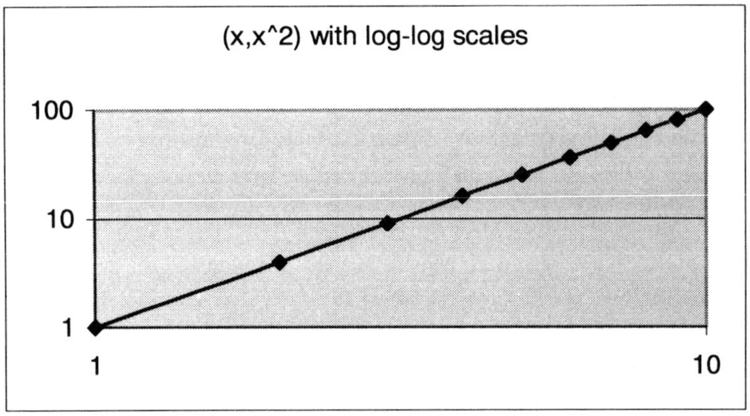

4. The graph of x1t = log x and y1t = log x^3 =3 log x or eliminating the log x: y = 3x

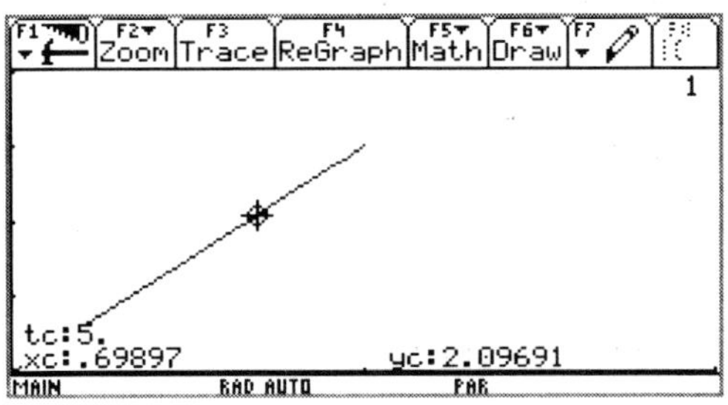

5a. 3, 4

5b. As in 4 above since log x^3 =3 log x, y = 3x a line with a slope of 3.

6a.

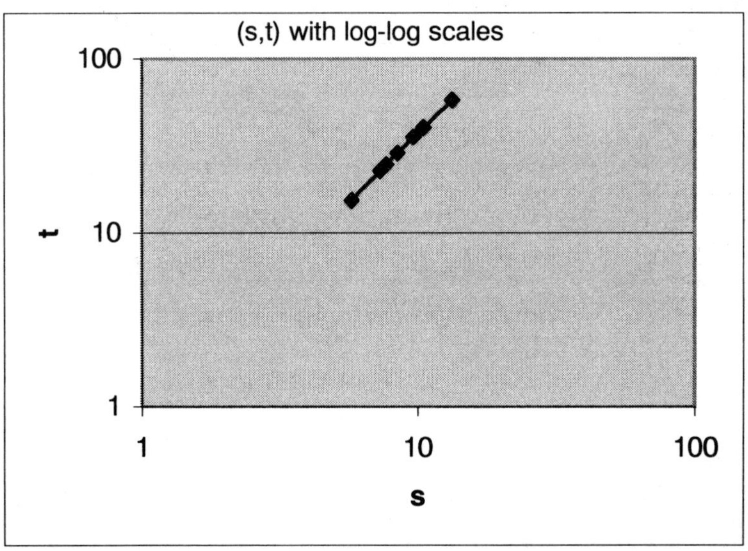

6b.

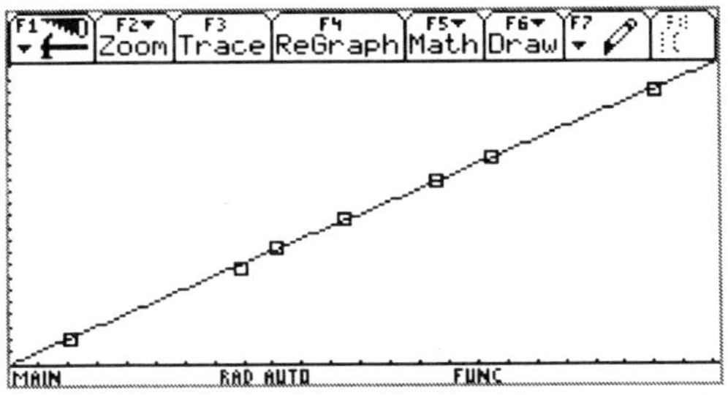

The data (log s, log t) with the regression line
t = 1.57s-0.000113 which rounded slightly is
log t= 1.57(log s) = log $s^{1.57}$

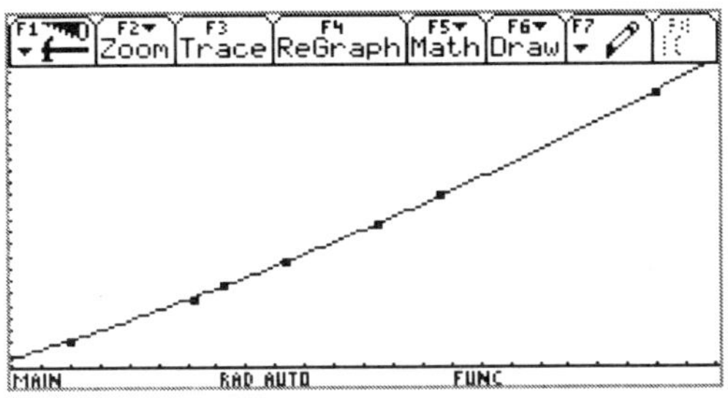

The data (s,t) with the power regression
t = $0.9997s^{1.57}$ again with a little rounding
t= $s^{1.57}$ compare this with the answer to part 6.

Exploration 32: Speedy Pistons

1. Stroke = 3 inches,
2. OC = 1.5, CP = 5
   a. and b.

$$CP^2 = OC^2 + OP^2 - 2(OP)(OC)\cos\alpha \qquad OP = \frac{3\cos\alpha + \sqrt{9\cos^2\alpha + 91}}{2}$$

4. a, b, c and 6a. 7 in BOLD and 8

Angle	Dist from Center	Difference	Difference $\div \frac{\pi}{20}$)	100 rpm Inches per min	3000 rpm Inches per min	Miles per hour	15,000 rpm inches per min	Miles per hour
0	6.5	---	---					
0.15708	6.476023	-0.02398	-0.1526402	-610.56	-18316.8	-17.35	-91584.1	-86.73
0.314159	6.405053	-0.07097	-0.4518123	-1807.25	-54217.5	-51.34	-271087.4	-256.71
0.471239	6.289919	-0.11513	-0.73296753	-2931.87	-87956.1	-83.29	-439780.5	-416.46
0.628319	6.135176	-0.15474	-0.98512143	-3940.49	-118214.6	-111.95	-591072.9	-559.73
0.785398	5.946865	-0.18831	-1.19882376	-4795.30	-143858.9	-136.23	-719294.3	-681.15
0.942478	5.732178	-0.21469	-1.36673852	-5466.95	-164008.6	-155.31	-820043.1	-776.56
1.099557	5.49905	-0.23313	-1.48414288	-5936.57	-178097.1	-168.65	-890485.7	-843.27
1.256637	5.255692	-0.24336	-1.54926677	-6197.07	-185912.0	-176.05	-929560.1	-880.27
1.413717	5.010116	-0.24558	-1.56338133	**-6253.53**	**-187605.8**	**-177.66**	**-938028.8**	**-888.29**
1.570796	4.769696	-0.24042	-1.53056262	-6122.25	-183667.5	-173.93	-918337.6	-869.64
1.727876	4.540813	-0.22888	-1.45711558	-5828.46	-174853.9	-165.58	-874269.4	-827.91
1.884956	4.328641	-0.21217	-1.35073032	-5402.92	-162087.6	-153.49	-810438.2	-767.46
2.042035	4.137078	-0.19156	-1.21952323	-4878.09	-146342.8	-138.58	-731713.9	-692.91
2.199115	3.968823	-0.16826	-1.07114867	-4284.59	-128537.8	-121.72	-642689.2	-608.61
		. . .		. . .				
4.39823	4.328641	0.191562	1.21952323	4878.09	146342.8	138.58	731713.9	692.91
4.555309	4.540813	0.212172	1.35073032	5402.92	162087.6	153.49	810438.2	767.46
4.712389	4.769696	0.228883	1.45711558	5828.46	174853.9	165.58	874269.4	827.91
4.869469	5.010116	0.24042	1.53056262	6122.25	183667.5	173.93	918337.6	869.64
5.026548	5.255692	0.245575	1.56338133	**6253.53**	**187605.8**	**177.66**	**938028.8**	**888.29**
5.183628	5.49905	0.243358	1.54926677	6197.07	185912.0	176.05	929560.1	880.27
5.340708	5.732178	0.233129	1.48414288	5936.57	178097.1	168.65	890485.7	843.27
5.497787	5.946865	0.214687	1.36673852	5466.95	164008.6	155.31	820043.1	776.56
5.654867	6.135176	0.188311	1.19882376	4795.30	143858.9	136.23	719294.3	681.15
5.811946	6.289919	0.154743	0.98512143	3940.49	118214.6	111.95	591072.9	559.73
5.969026	6.405053	0.115134	0.73296753	2931.87	87956.1	83.29	439780.5	416.46
6.126106	6.476023	0.070971	0.4518123	1807.25	54217.5	51.34	271087.4	256.71
6.283185	6.5	0.023977	0.1526402	610.56	18316.8	17.35	91584.1	86.73

Note : these figures assume a 3 inch stroke. Many engines, especially race engines have a much smaller stroke, meaning a lower piston speed. Find out a more realistic value for the stroke and refigure the speeds.

4d. velocity is not constant.
4e. When the piston moves up the velocity is positive, moving down the velocity is negative.

Exploration 33: Moving Forward By Sitting Still

1. The track is 44.188 ft long.
2. It completes a full cycle in 6.0257 seconds The linear velocity of the bulldozer is 7.3333 ft per sec (5 mph).
3. The wheels are turning at 11 radians per second.
4.

Portion of the cycle	Time for this portion	Time Running total
Stationary on the ground	2.7272	2.7272
Up the back wheel	.285599	3.01287
Across the top	2.7272	5.74014
Down the front wheel	.285599	6.0257

5.

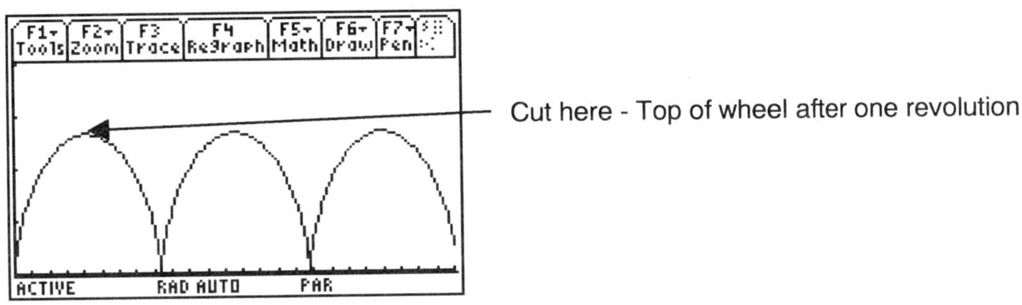

6.
x1t(t) = 0 y1t(t) = 0 $0 \le t < 2.7272$

x2t(t) = .6667 ((t - 2.7272)*11 – sin ((t – 2.7272)*11))
y2t(t) = .6667(1 – cos ((t – 2.7171)*11)) $2.7272 \le t < 3.01287$

x3t(t) = .6667(π - sin π) + (t – 3.01287)*7.333
y3(t) = 1.3333 $3.01287 \le t \le 5.74014$

x4t(t) =20 + .6667 ((t – 5.74014)*11 – sin ((t – 5.74014)*11))
y4t(t) = .6667 (1 – cos((t – 5.74014)*11)) $5.74014 \le t \le 6.0257$

(Note: In x4t(t), 20 = x3t(5.74014) the starting x-position for this part of the path)

7. Note for the first part of the trip the point on the track does not move; it remains at (0,0). The curve at the left is the second part (up the back wheel), the long segment is the third part (across the top) and the curve at the right is the last part (down the front wheel).

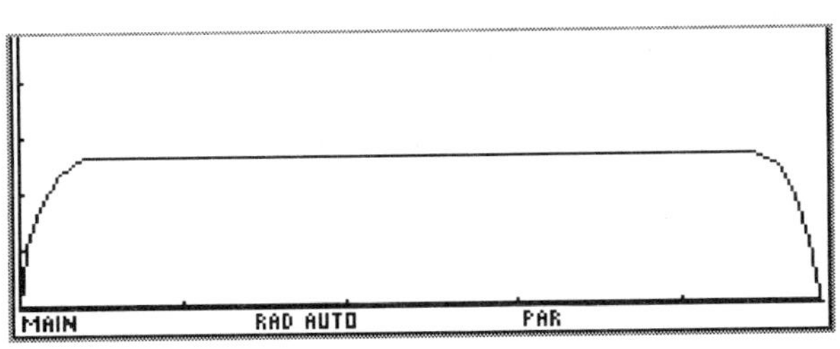

8. Extension

Time	0	.0238	.0476	.0714	.0952	.119	.1428	.1666	.1904	.2142	.238	.2618
Position	0	.00199	.01573	.0522	.12079	.2287	.3806	.57781	.81895	1.0994	1.412	1.747
Difference	*	.00199	.01375	.03646	.06859	.1079	.1518	.19726	.24115	.28049	.3126	.3353
Velocity		0.0835	0.5776	1.5321	2.8820	4.5352	6.3792	8.2882	10.132	11.785	13.135	14.090

b. The velocity plotted as data

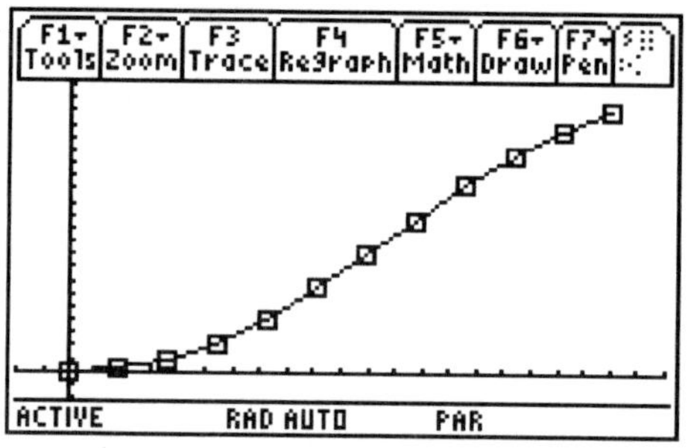

c. The corresponding portion of the position graph (second part of path – up the back wheel). The x-coordinate increases slowly at first, then faster as it approaches the top of the wheel.

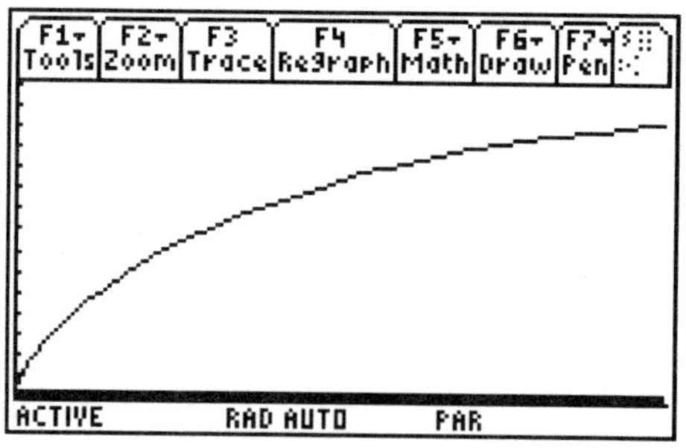

Chapter 5

Exploration 34: Fidget With Widgets

1.

$$\text{Cost}(x) = \begin{cases} 23.33 * x \text{ for } x < 20 \\ 18.75 * x \text{ for } 20 \le x < 50 \\ 12.25 * x \text{ for } x \ge 50 \end{cases}$$

2. Cost (light graph); Income (heavy graph). The window is [0,70] by [0,1500]

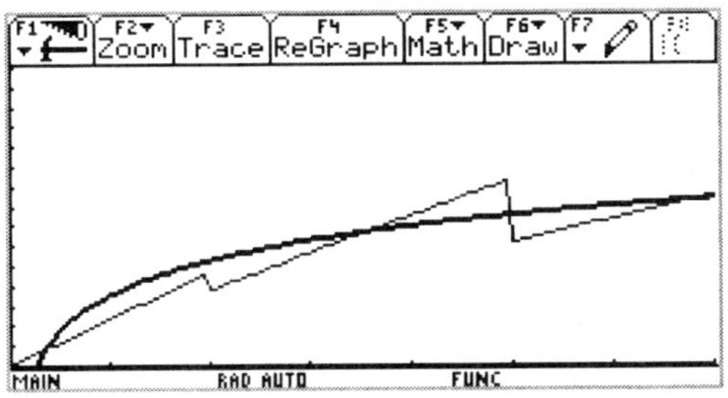

3. Profit = Income(x) – Cost (x)

4. Graph Profit is the heavy graph. The maximum is at 11 items.

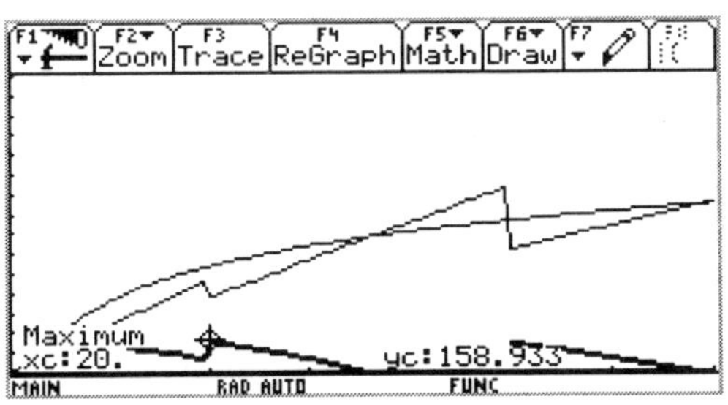

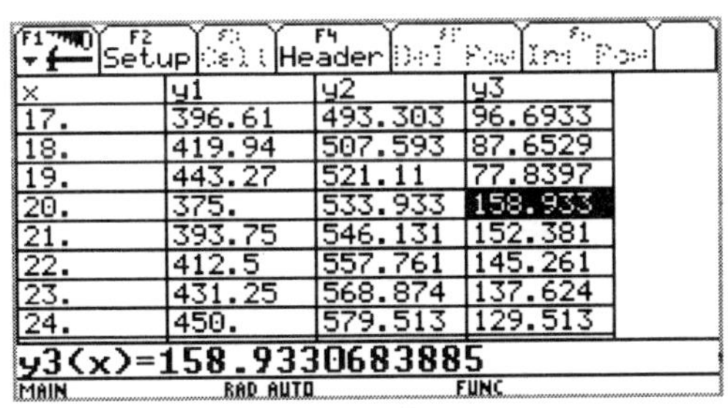

x	y1	y2	y3
17.	396.61	493.303	96.6933
18.	419.94	507.593	87.6529
19.	443.27	521.11	77.8397
20.	375.	533.933	158.933
21.	393.75	546.131	152.381
22.	412.5	557.761	145.261
23.	431.25	568.874	137.624
24.	450.	579.513	129.513

y3(x)=158.9330683885

Exploration 35: Breaking Everything or Going Broke

1. Cost and Sales graphs:

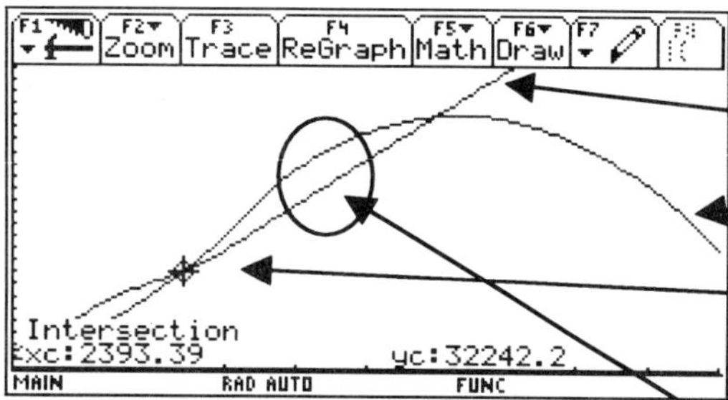

2. $29,675
3. $26,998.78
5. Profit = income – cost

6. Profit graph:

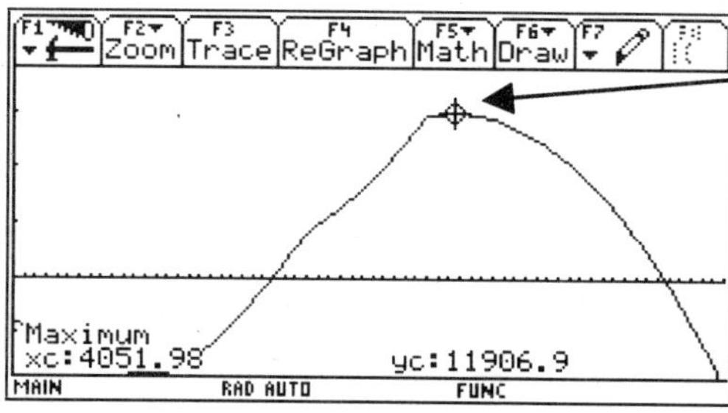

8a. 4,052 items
8b. $11,901.80

Exploration 36: Payoff

1. Income function:

$$I(d) = \begin{cases} 20 + 5515*d - 128*d^2 & \text{for } d < 30 \\ 50000 & \text{for } d = 30 \\ 50000-1000(x-30) & \text{for } d > 30 \end{cases}$$

2. Payoff: Window [0, 40] by [0, 60,000]

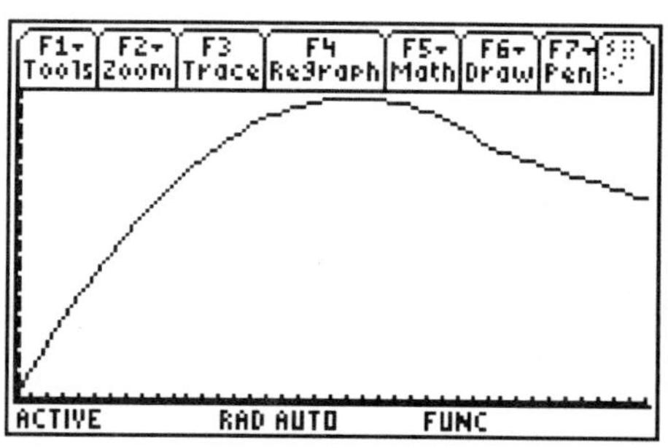

3. Daily_labor_cost (m) = $\begin{cases} 12.5m & m \le 80 \\ 80(12.5) + 18.75(m - 80) & m > 80 \end{cases}$

where m is the man hours worked.

4. m = Man_hours (d) = $\dfrac{3000}{d}$ with d in days worked

Labor_cost (d) = Daily_labor_cost (Man_hours(d)) * d

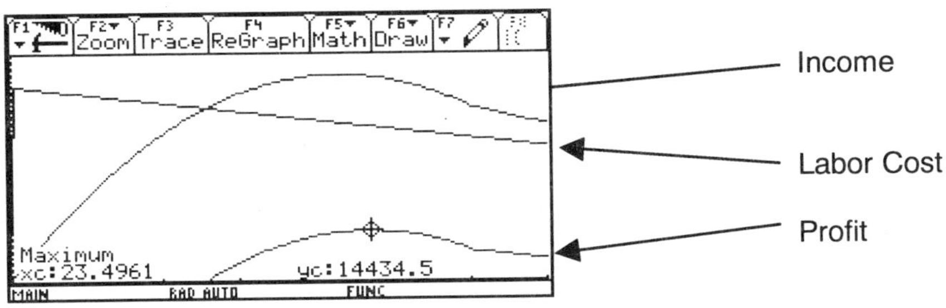

Exploration 37: Time to Re-Tire?

1. Rate = 180 + 1.5x (x = added % of SX47) in mph
2. Dist = 160 – 20x
3. Number of tire changes = int (500/(160-20x)+1)
4. Time for tire changes = int ((500/(160-20x))+1) * 0.025 (# changes * 90 sec)
5. Time = (500/rate) + time for tire changes

6.

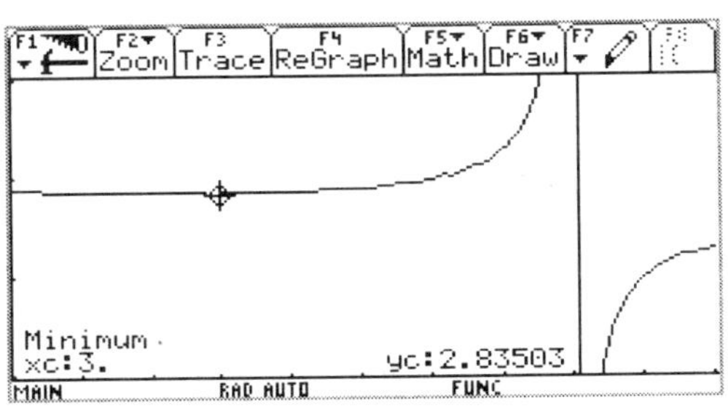

7. Min time at x = 3% additional SX47, 2.835 hours for 500 miles with 5 tire changes

Exploration 38: Timberrrr!

1. The radius is 4 inches and the length must be changed to inches, so the income fro each tree is $\dfrac{\pi * 4^2 * 40 * 12}{144} * 0.10$ dollars

2. Net each tree = $\dfrac{\pi * 4^2 * 40 * 12}{144} * 0.10 - 5 = 11.7552$

3. The final function is $11.7552x - (.27147x^2 - 39.0925x)$

4.

5. The optimum number of trees is 93 – 94 with a profit of about $2380 for an acre

Exploration 39: Putting Customers in Satellite Orbit

1. Number of customers anticipated = 1500 + 5x
2. Yearly income generated = (1500 + 5x)(700-x)

3. Number of customers willing to pay extra = $(7500+1500+5x)* .01*\frac{700}{\sqrt{3000-x}}$

4. Yearly income generated by extra channel purchases =

$$12*(.10)*x * (7500 + 1500+5x)* .01*\frac{700}{\sqrt{3000-x}}$$

5. Total income generated is the sum of the income from initial set up and monthly fees (equation 2 above plus equation 4)

6. Optimum price reduction = $437.89 for income of $1.78 million

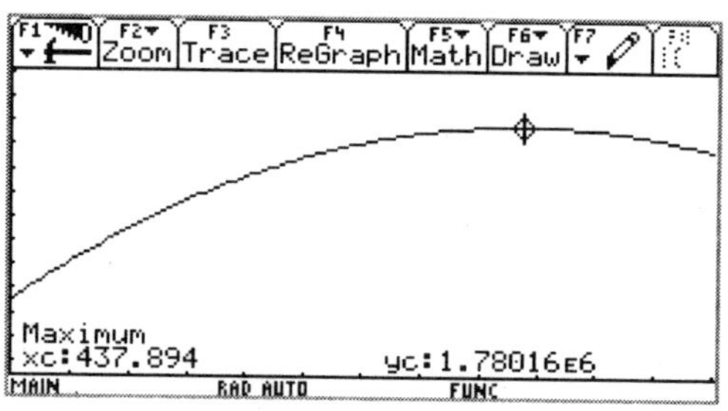

Exploration 40: Hang On to That Paint Bucket!

1. Daily Attendance in year x = $12000*1.015^{x}$
2. Ticket cost in year x = 23.50 + 0.25x
3. Daily Revenue in year x = $(12000*1.015^{x})*(23.50+0.25x)$

90 day loss during year x = $(12000*1.015^{x})*(23.50+0.25x)*90*0.10$

4. Paint cost for x extra years = 10000*x
5.

Year	Loss if shut down this year	Difference in loss if wait another year	Additive cost for delaying painting until this year
5	$ 2,879,580		$ -
6	$ 2,952,297	$ 72,717	$ 10,000
7	$ 3,026,547	$ 74,250	$ 20,000
8	$ 3,102,361	$ 75,814	$ 30,000
9	$ 3,179,768	$ 77,407	$ 40,000
10	$ 3,258,799	$ 79,031	$ 50,000
11	$ 3,339,485	$ 80,687	$ 60,000
12	$ 3,421,859	$ 82,374	$ 70,000
13	$ 3,505,953	$ 84,094	$ 80,000
14	$ 3,591,800	$ 85,847	$ 90,000
15	$ 3,679,433	$ 87,633	$ 100,000
16	$ 3,768,887	$ 89,454	$ 110,000
17	$ 3,860,197	$ 91,310	$ 120,000
18	$ 3,953,398	$ 93,201	$ 130,000
19	$ 4,048,527	$ 95,129	$ 140,000

6. Until year 13 the cost of the better paint is more than the money lost by waiting.

7.

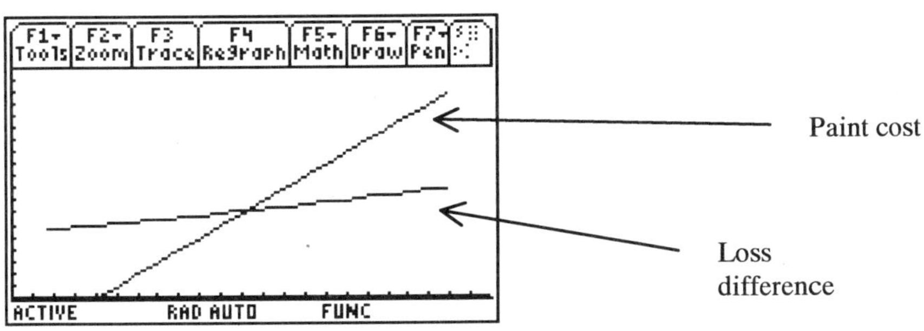

Exploration 41: Go Fish!

1. Spoilage_percent (t) = 4*(t-32)*.03 (as a decimal)

2. Value (t) = (1-4*(t-32)*.03)*4000*1.57

3. Profit (t) = $Value(t) - 4 * \sum_{i=t}^{40} 50 + (40 - i) * 20$

4.

Temp	Cooling Cost	Value	Profit
32	$ 4,680	$6,280.00	$1,600.00
33	$ 3,840	$5,526.40	$1,686.40
34	$ 3,080	$4,772.80	**$1,692.80** ←
35	$ 2,400	$4,019.20	$1,619.20
36	$ 1,800	$3,265.60	$1,465.60
37	$ 1,280	$2,512.00	$1,232.00
38	$ 840	$1,758.40	$ 918.40
39	$ 480	$1,004.80	$ 524.80
40	$ 200	$ 251.20	$ 51.20

5. Maximum profit at 34 degrees

6.

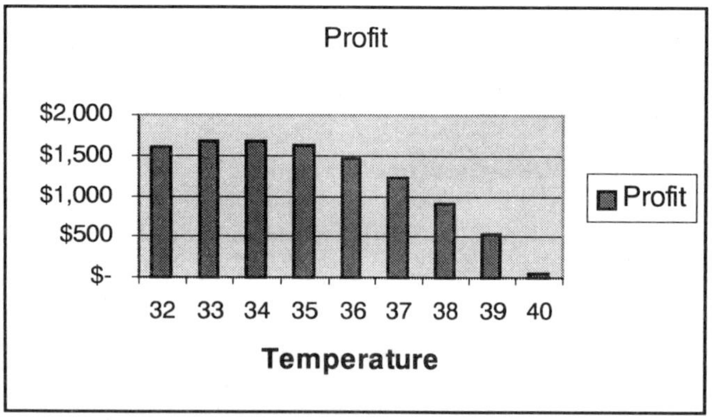

Exploration 42: Beam Me Up

2. Height = 3 * depth

3. Width = $\sqrt{36^2 - (3d)^2}$

4. depth^2 * width * = $d^2\sqrt{36^2 - (3d)^2}$

depth = 9.7979; width = 20.785

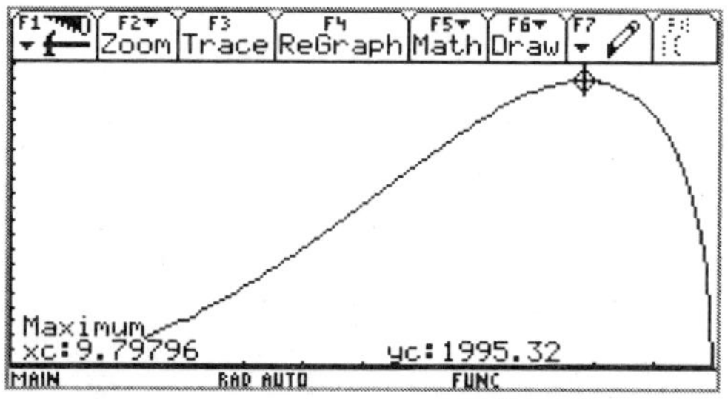

5.

6. Waste = $\pi 18^2 - 3(9.7879)(20.785) = 407.55$ square inches wasted.

7. depth * width^2 = $d * (36^2 - (3d)^2)$

depth = 6.9282; width = 29.3939

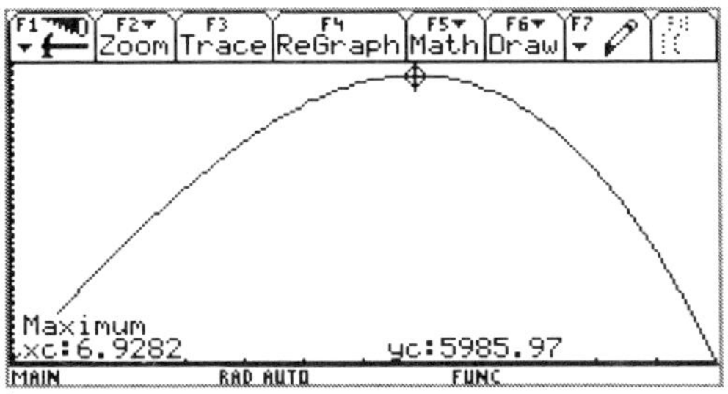

Total cross section for three beams in either case is the same = 610.948 square inches. Waste is the same for either configuration. Issue now is which dimension is the stronger beam.

8. The second way is stronger. The strength is given by the y-coordinate of the maximum point. The strength in the first case is 1995 and in the second 5985,97.

Exploration 43: Lighter Is Faster

1. $V_t(m) = 190 - 0.3325*m$
2. $V_g = 159 + 9.06 * \ln(.4 * m)$
3. Graph I [0,50] by [150,200]

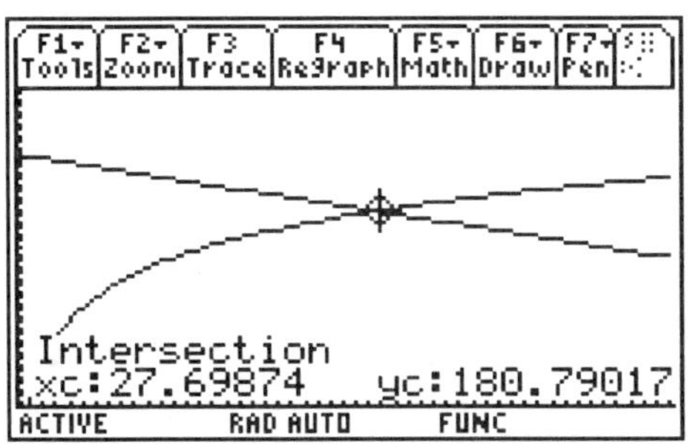

4. and 6 *m* = 27.7 is the number of miles after starting with a new ties and a full tank of gas, when the car can go the fastest. This is the best time to pass.
5.

Miles	Tire Speed	Gas Speed
5	188.34	165.28
10	186.68	171.56
15	185.01	175.23
20	183.35	177.84
25	181.69	179.86
30	180.03	181.51
35	178.36	182.91
40	176.70	184.12
45	175.04	185.19
50	173.38	186.14

Exploration 44: Trash or Treasure?

1. 169,333 cubic yards
2. 339 months (28.22 years) to fill the pit
3. V(t) = 500* t
4. 500*t = 169333 when t = 339.667 months

5.

Months	% Reduction of Volume as a function of time	Volume (in 1000s yd^3)	% Reduction of Volume
1	0.004500	10	0.1300
5	0.005916	50	0.1993
10	0.006656	100	0.2291
20	0.007488	160	0.2493

6a. First 12 months of composting:

Month	Refuse Added	% CTR(t) as a decimal	%VTR(v) as a decimal	Amount
0	500	---------	---------	10000
1	500	0.000045	0.00130045	10487
2	500	0.000051	0.00132087	10972
3	500	0.000054	0.00134034	11457
4	500	0.000057	0.00135893	11941
5	500	0.000059	0.00137671	12423
6	500	0.000061	0.00139376	12905
7	500	0.000063	0.00141012	13386
8	500	0.000064	0.00142586	13866
9	500	0.000065	0.00144101	14346
10	500	0.000067	0.00145561	14824
11	500	0.000068	0.00146971	15301
12	500	0.000069	0.00148334	15777

6b. $V(t+1) = V(t) + 500 - 0.0045t^{0.17} \cdot V(t) + (0.266 - .043\ln(V(t))) \cdot V(t)$

simplifies to $-.011438 \cdot V(t) \cdot \ln(V(t)) + V(t) \cdot (1 - .0045 \cdot t^{.17}) + 500$

6c. Last 6 months when landfill becomes full.

776	500	0.000139	0.00251627	169080
777	500	0.000140	0.00251640	169131
778	500	0.000140	0.00251652	169182
779	500	0.000140	0.00251665	169233
780	500	0.000140	0.00251678	169283
781	500	0.000140	0.00251691	169333

6d. Increases life of the landfill by 462 months (already 20 months when composting started) or 38.5 years.

Exploration 45: Free or Best Offer!

1. Rebate_percent (a) = .00001 * a (a linear function)
   (up to $100,000)

2.

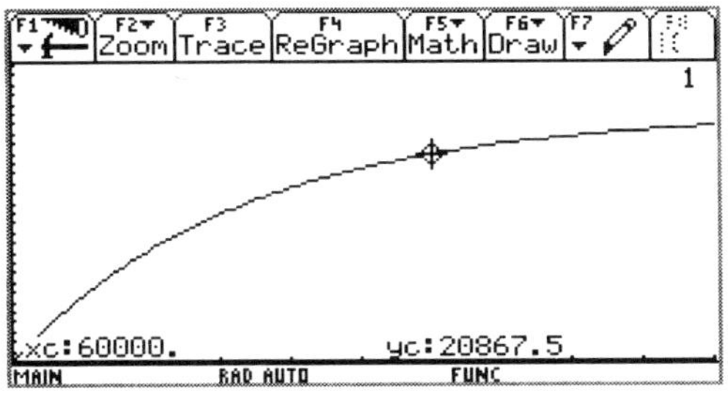

Number of sales as a function of advertising dollars

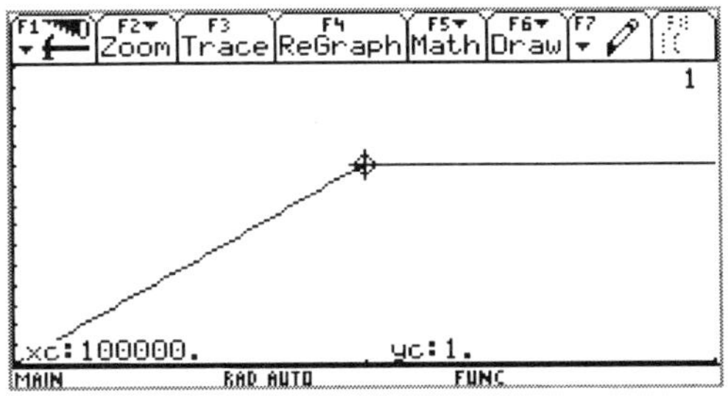

Percent (in decimal form) of sales who are rebate users as a function of advertising dollars

3. Income(a) = 300 * (25000 – 25000e$^{-0.0003a}$) - a - 150(0.00001*a)*(25000 - 25000 e$^{-0.0003a}$)

Graph of Income function with maximum:

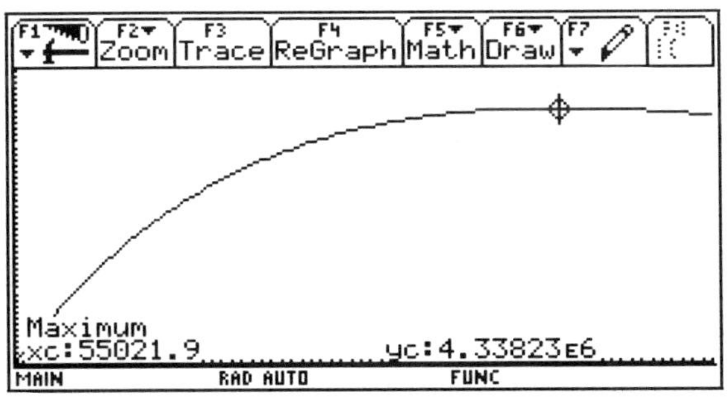

Chapter 6

Exploration 46: Build a Better Balloon

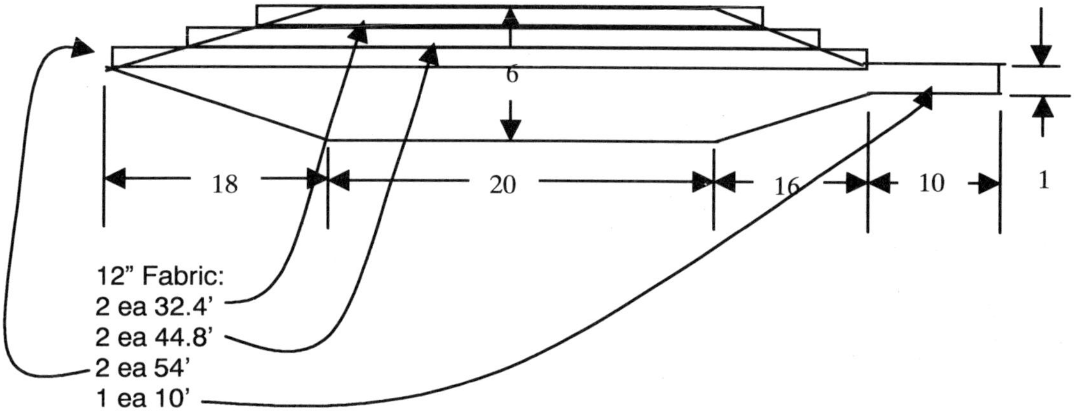

272.4 ft x 16 = 4358.4 ft =1452.8 yards => $2542.4
Seams: 2 x 32.4 + 2 x 44.8 + 1 x 54 + 1 = 209.4 ft x 16 = 3350.4 ft => $67.01
Total Cost $2609.41

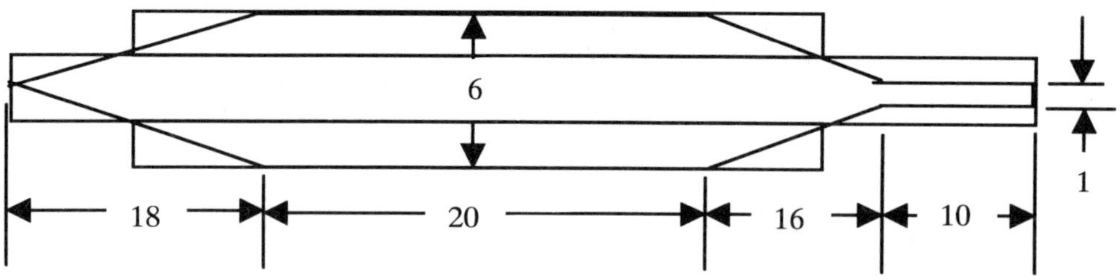

24" Fabric:
2 ea 41.8'
1 ea 64'

Fabric 147.6 x 16 = 2361.6 feet => 787.2 yds => $2361.60
Seams: 2 ea 41.8 = 83.6 ft x 16 = 1337.6 ft => $26.75 for seams
Total Cost = $2388.35
36" Fabric:
2 ea 64' = 128 ft x 16 = 2048 ft = 682.667 yds => $2901.33 for fabric
Seams: 1 ea 64 ft x 16 = 1024 ft => $20.48
Total Cost $2921.81

72" Fabric
1 ea 64' x 16 = 1024 ft => 341.33 yds => $1877.00 for fabric, no seams

There are lots more possibilities when you start allowing other combinations of patterns. This is one of those problems where you let the students' creativeness and problem-solving skills run wild to try to puzzle together a better solution, if possible

Exploration 47: The Stock Market – Hindsight is 20/20

1a. Both bogus Box and Snidely are double their original price at the end of the chart. Each will make you a profit of $1,000
1b. Snidley quadruples in price between Jan. 1995 and April 1995. You make a profit of $3000.
1d.Either with the table or graph look for the largest positive change from beginning to end.
1e. Either with the table or graph look for the largest positive change in the shortest time. The graph below shows the percent change. This may also help.

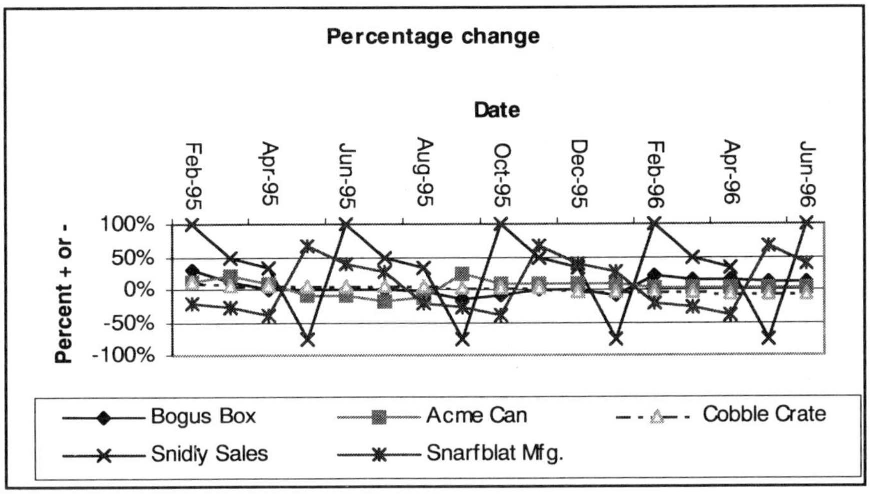

2a. Buy more shares of the cheaper stock.
2b. The percent increase
2c. The slope is the percent change expressed as a decimal
2d. The steepest positive slope is the largest change.
3. Possible sequence of purchases. You might challenge students before the exercise to estimate the order of magnitude of the results. 10's of thousands? 100's of thousands? millions? 10's of millions?

Date	Stock Chosen	Price/Share	# Shares Purchased	Value at end of month
Jan-95	Snidly	10	100	2,000
Feb-95	Snidly	20		3,000
Mar-95	Snidly	30		4,000
Apr-95	Snarfblat	30	133.33	6,666.67
May-95	Snidly	10	666.67	13,333.40
Jun-95	Snidly	20		20,000.10
Jul-95	Snidly	30		26,666.67
Aug-95	Acme	80	333.34	33,333.50
Sep-95	Snidly	10	3,333.35	66,667.00
Oct-95	Snarfblat	30	2,222.23	111,111.67
Nov-95	Snarfblat	50		7,777,816.67
Dec-95	Snarfblat	70		700,003,500
Jan-96	Snidly	10	70,000,350	1,400,007,000
Feb-96	Snidly	20		42,000,210,000
Mar-96	Snidly	30		21,680,008,400,000
Apr-96	Snarfblat	30	56,000,280,000	84,000,420,000,000
May-96	Sindly	10	840,004,200,000	168,000,840,000,000